The Portuguese at War

The Portuguese-Speaking World

ITS HISTORY, POLITICS AND CULTURE

This series will publish high-quality scholarly books on the entire spectrum of the Portuguese-speaking world, with particular emphasis on the modern history, culture, and politics of Portugal, Brazil, and Africa. The series, which will be open to a variety of approaches, will offer fresh insights into a wide range of topics covering diverse historical and geographical contexts. Particular preferences will be given to books that reflect interdisciplinarity and innovative methodologies. The editors encourage the submission of proposals for single author as well as collective volumes.

PUBLISHED

The Lusophone World: The Evolution of Portuguese National Narratives
Sarah Ashby

The Politics of Representation: Elections and Parliamentarism in Portugal and Spain, 1875–1926
Edited by Pedro Tavares de Almeida & Javier Moreno Luzón

Inequality in the Portuguese-Speaking World: Global and Historical Perspectives
Edited by Francisco Bethencourt

Marcello Caetano: A Biography (1906–1980)
Francisco Carlos Palomanes Martinho

From Lisbon to the World: Fernando Pessoa's Enduring Literary Presence
George Monteiro

The First Portuguese Republic: Between Liberalism and Democracy (1910–1926)
Miriam Pereira

The Portuguese at War: From the Nineteenth Century to the Present Day
Nuno Severiano Teixeira

The Locusts: British Critics of Portugal before the First World War
Gary Thorn

FORTHCOMING

Politics and Religion in the Portuguese Colonial Empire in Africa (1890–1930)
Hugo Gonçalves Dores

The Military and Political in Authoritarian Brazil: The Aliança Renovadora Nacional (ARENA), 1965–1979
Lucia Grinberg

The Eruption of Insular Identities: A Comparative Study of Azorean and Cape Verdean Prose
Brianna Medeiros

Dictatorship and the Electoral Vote: Francoism and the Portuguese New State Regime in Comparative Perspective, 1945–1975
Carlos Domper Lasús

Literary Censorship in Francisco Franco's Spain and Getulio Vargas' Brazil, 1936–1945: Burning Books, Awarding Writers
Gabriela de Lima Grecco

On Guard Against the Red Menace: Anti-Communism in Brazil, 1917–1964
Rodrigo Patto Sá Motta

For my wife,
Cármen

The Portuguese at War

From the Nineteenth Century to the Present Day

NUNO SEVERIANO TEIXEIRA

SUSSEX
ACADEMIC
PRESS
Brighton • Chicago • Toronto

2 4 6 8 10 9 7 5 3

First published in hardcover 2019, reprinted in paperback 2020, in Great Britain by
SUSSEX ACADEMIC PRESS
PO Box 139
Eastbourne BN24 9BP

Distributed in North America by
SUSSEX ACADEMIC PRESS
Independent Publishers Group
814 N. Franklin Street
Chicago, IL 60610

British Library Cataloguing in Publication Data
A CIP catalogue record for this book is available from the British Library.

Library of Congress Cataloging-in-Publication Data
Names: Teixeira, Nuno Severiano, author.
Title: The Portuguese at war : from the nineteenth century to the present day / Nuno Severiano Teixeira.
Description: Brighton ; Chicago : Sussex Academic Press, [2019] | Series: The Portuguese-speaking world : its history, politics and culture | Includes bibliographical references and index.
Identifiers: LCCN 2019008400 | ISBN 9781845199937 (hbk : alk. paper) | ISBN 9781789760576 (pbk : alk. paper)
Subjects: LCSH: Portugal—History, Military.
Classification: LCC DP547 .T45 2019 | DDC 355.0209469—dc23
LC record available at https://lccn.loc.gov/2019008400

Typeset and designed by Sussex Academic Press, Brighton & Eastbourne.
Printed by Independent Publishers Group, Chicago

Contents

Series Editors' Preface ix
Author's Preface xi
Acknowledgements xiii
List of Abbreviations xiv

1
Between Liberal Revolution and Civil Wars 1
The international context of liberal revolution
Revolution and counter-revolution: Absolutism, liberalism and civil war
Two liberalisms in conflict: Constitution, chart and civil war

2
Building the African Empire 22
The new Brazil
The scramble for Africa and effective occupation
Metropolitan army and colonial army
"Pacification" campaigns

3
The Republic and the Great War 47
The international standing of the Republic before the war
Entering the war: National objectives and political strategies
The conduct of war: War policies, the military instrument and the operational theatres
Managing peace: The aftermath of war and the Versailles Conference

4
The *Estado Novo* and the Spanish Civil War 77
The international standing of the *Estado Novo* before the war
Pushing the limits of ambiguity: Portugal in the face of the Spanish civil war

From the internal threat to the continental threat: Defence policy and military reforms

5
The *Estado Novo* and the Second World War 92
Portugal's international standing and foreign policy on the eve of the Second World War
Why neutrality? Portugal and the Second World War
The avatars of continental threat: Defence policy, strategy and the military instrument

6
The Cold War and NATO 110
The international standing of the *Estado Novo* and the second post-war period
From neutrality to alignment: Portugal and the foundation of the Atlantic Alliance
Portugal and NATO: Defence policy and the reform of the military instrument

7
The End of Empire and the Decolonisation Wars 127
The international isolation of the *Estado Novo*: The international scene, Portuguese foreign policy and the colonial question
Mobilising for war: Defence policy, the armed forces and guerrilla warfare
Independence movements and armed conflict: One or three decolonisation wars?
War assessment and the political fate of the regime

8
From War Campaigns to Peacekeeping Operations 162
Portuguese Democracy between Africa and Europe
Post Cold War: Atlanticism, europeanisation and post-colonial relations
The democratic model of the armed forces and the new international missions.

Notes 194
General and Specialised Bibliography 222
Index 239

Series Editors' Preface

Nuno Severiano Teixeira's *The Portuguese at War* is a masterful synthesis of the centrality of war ethos and practice in the history of Portugal since the beginning of the nineteenth century, encompassing the most significant events of the liberal revolution (1820) and the ensuing civil wars. It provides a long overdue assessment of how the Portuguese military institutions viewed their responsibilities in multiple arenas, and informs of the debates and personalities involved.

As the author states clearly, this work does not aim to be "an exercise in historiographical controversy." That said, it nevertheless sets out with some precision the international and domestic issues pertaining to each different war footing scenario, exploring and articulating the circumstances and potentialities, and of course the dangers in the actions approved and acted upon. A particular strength of the argument presented is the ability to interconnect micro and macro dimensions; decision-making in Portugal often fell foul of local circumstance.

The book does not engage in a narrow "national"(ist) approach, since the issues are always analyzed in their wider framework. As distinct historical periods are discussed, geographies vary – from domestic to foreign, and of course Portugal's substantive colonial experience. In this latter respect, the author deals with the euphemistically named "pacification campaigns" (that gave a modicum of reality to the otherwise feeble colonial rule in the late nineteenth-century), to the violent imperial and colonial disengagements of the 1970s, themselves intimately associated with political-military dynamics at home, which ended soon after the Carnation Revolution of 1974.

The historical dynamics and contexts covered by *The Portuguese at War* include more contemporary, post-colonial and post-authoritarian momentums, which have brought about new challenges to the military establishment as well as to those in charge of its political administration. The Portuguese integration in the European Union simply reinforced these changes, along with the challenges of new post-cold war international sensibilities. This book's robust and multi-faceted analysis will no doubt become a significant reference to all those interested in understanding how ideas about war, and practices of war, shaped Portuguese history, particularly its once considerable overseas, colonial empire.

António Costa Pinto (University of Lisbon)
Onésimo T. Almeida (Brown University)
Miguel Bandeira Jerónimo (University of Coimbra)

Author's Preface

From war campaigns to peacekeeping operations, *The Portuguese at War* attempts to outline a panorama of Portugal's military history from the 19th century until the present day.

From the French invasions to the civil wars, from the African Empire to the wars of decolonisation, from intervention in the First World War to neutrality in the Second, and participation in international peace missions in the scope of NATO, the European Union and the United Nations. But also the revolutions and the civil wars: from the dynastic struggles to the liberal wars, from the implantation of the republic to the military dictatorship, from the authoritarian regime to the 25th of April and the transition to democracy.

It offers, therefore, a historical perspective of war and the military institution in Portugal – of the military recruitment systems and the social composition and organisation of the Portuguese Armed Forces: the Army, the Navy and, more recently, the Air Force; of the military *ethos*; of the evolution in equipment, armament and military technology; of the art of war and strategic thought; of tactics and military operations; of the fight and the fighters. This work is not merely about the grammar of war, however; it deals with war in its context: the international and European environment, and its relations with economy and society and with public opinion and political power.

As a synthesis, it does not – indeed it could not – dwell on every aspect with the same degree of detail and thoroughness. It strives, instead, to present a historical overview and a comprehensive insight into its subject.

As a work of synthesis, it seeks to balance accuracy and academic standards with a literary narrative style appealing not only to scholars, but also to a broader public audience. As such, and never losing sight of the plurality of interpretative theories, it is not meant as an exercise in historiographical controversy.

The book is chronologically organised into eight chapters. Each chapter addresses a conflict in its relation with the international strategic environment and the domestic political regime. The first chapter deals with the liberal revolution and the civil wars under the constitutional monarchy in the first half of the 19th century. The second tackles the military campaigns in the African empire in the second half of the 19th century, still under the monarchic rule. The following chapters deal with the 20th century. The third is devoted to Portuguese military intervention in the Great War, already under the republican regime. The fourth and fifth, on Spain's civil war and the Second World War, under Salazar's regime, still focussing on the first half of the 20th century. The sixth and the seventh address the Cold War and the decolonisation wars, still under Salazar's regime, in the second half of the 20th century. Lastly, the eighth chapter addresses the democratic regime from the last quarter of the 20th to the first decade of the 21st century: the revolution of the 25th of April and the participation of the Portuguese Armed Forces in peacekeeping operations.

The bibliography provided is not and could never be exhaustive, comprehending the whole documental corpus that supported a work of synthesis extending for more than two centuries. It is split into general bibliographic entries and specialised entries; its purpose is to work as a reading guide for those who wish to delve further into the subject.

For readers' ease of reference, a comprehensive index has been compiled.

Acknowledgements

I would like to express my gratitude to my host institutions, which have provided the institutional and financial conditions essential for the publication of this book: to the NOVA University of Lisbon, for granting me my sabbatical leave; to Georgetown University, wherein, as Visiting Professor, the present work was carried out; and to the Portuguese Institute of International Relations, without whose financial support this book would have never seen the light of day.

A special note of thanks to Sara Fevereiro, for her high degree of professionalism and unwavering congeniality in the arduous translation work.

Lastly, "thank you" is not really enough for the invaluable support given by those close to me. To my family, for the many stolen hours, and especially to Cármen, to whom I dedicate this book.

List of Abbreviations

CEP	Portuguese Expeditionary Corps
CFSP	Common Foreign and Security Policy
CIA	Central Intelligence Agency
CINCIBERLANT	Commander-in-Chief Iberian Atlantic Area
CINCSOUTHLANT	Commander-in-Chief Southern Atlantic
CJTP	Combined Joint Task Force
CLSTP	Committee for the Liberation of São Tomé e Príncipe
CONCP	Conference of Nationalist Organisations of the Portuguese Colonies
COREMO	Mozambique Revolutionary Committee
CPLP	Community of Portuguese-Speaking Countries
CSDP	Common Security and Defence Policy
EASTLANT	Eastern Atlantic Area
EEC	European Economic Union
ESDI	European Security and Defence Identity
ESDP	European Security and Defence Policy
EU	European Union
EUFOR (ALTHEA)	European Union Force
EUROFOR	European Rapid Operational Force
EUROMARFOR	European Maritime Force
FLAD	Luso-American Development Foundation
FLEC	Front for the Liberation of the Enclave of Cabinda
FNLA	National Liberation Front of Angola
FRAIN	African Revolutionary Front for the National Independence of the Portuguese Colonies
FRELIMO	Liberation Front of Mozambique
GNR	National Republican Guard
GRAE	Revolutionary Government of Angola in Exile
IBERLANT	Iberian Atlantic Area
ICAUNC	Action Committee of the Cabinda National Union
INTERFET	Multinational Force for East Timor

ISAF	International Security Assistance Force
KFOR	Kosovo Force
MAC	Anti-Colonial Movement
MANU	Mozambique African National Union
MDN	Ministry of National Defence
MFA	Armed Forces Movement
MIA	Movement for the Independence of Angola
MING	Movement for the National Independence of Guinea
MINURSO	United Nations Mission for the Referendum in Western Sahara
MLEC	Movement for the Liberation of the Enclave of Cabinda
MONUA	United Nations Observer Mission in Angola
MPLA	People?s Movement for the Liberation of Angola
NATO	North Atlantic Treaty Organisation
NRF	NATO Response Force
OAU	Organisation of African Unity
OMLT	Orientation Mentor Liaison Team
OSCE	Organization for Security and Co-operation in Europe
PAI	African Party for the Independence
PAIGC	Guinea and Cape Verde Independence Party
PALOP	Portuguese-Speaking African countries
PCA	Angolan Communist Party
PCP	Portuguese Communist Party
PLUA	Party of the United Struggle for Africans in Angola
PVDE	State Surveillance and Defence Policy
RRF	Rapid Reaction Force
SACEUR	Supreme Allied Commander Europe
SACLAND	Supreme Allied Commander Atlantic
SFOR	The Stabilisation Force in Bosnia and Herzegovina
STANAVFORLANT	Standing Naval Force Atlantic
STRIKFORNATO	Naval Striking and Support Forces NATO
UDENAMO	National Democratic Movement of Mozambique
UN	United Nations
UNAMI	National African Union of Independent Mozambique
UNAVEM	UN Angola Verification Mission

UNESCO	United Nations Educational, Scientific and Cultural Organization
UNIFIL	United Nations Interim Force in Lebanon
UNITA	National Union for the Total Independence of Angola
UNOGIL	United Nations Observation Group in Lebanon
UNOMOZ	United Nations Operations in Mozambique
UNTAET	United Nations Transnational Administration for East-Timor
UNTAG	United Nations Transition Assistance Group in Namibia
UPA	Union of the Peoples of Angola
UPNA	Union of the Peoples of Northern Angola
WESTLANT	Western Atlantic Area
WEU	Western European Union

The Portuguese at War

CHAPTER

1

Between Liberal Revolution and Civil Wars

The international context of the liberal revolution

The breakdown of the Old Regime and the implantation of Liberalism was a slow, intricate process marked by revolutions and civil wars which spread across the first half of the 19th century, between 1820 and 1851. Throughout that process, the military invariably played a central role.

Ever since the 19th century, the ideals of the French Revolution had been flourishing in Portugal, eliciting a backlash reaction from the traditionalist forces of Throne and Altar. The French invasions had a huge, if ambiguous, impact on this movement: on the one hand, they put the country in touch with the ideal and experience of Napoleonic France; the other hand, the violence of the invasions sparked a nationalist reaction against the French and all that pertained to them. The invasions had one more outcome: the peninsular war between France and England resulted in a strategic unification of the Iberian Peninsula, which became a single theatre of operations. Battles took place, tactically, in one or the other – Portuguese or Spanish – side of the border, but their strategic purpose went far beyond either one of them. As a consequence, the liberal wars that ensued in Portugal and Spain were marked by mutual influence and support at a political and military level.

One thing is certain: the implantation of Liberalism in Portugal was intrinsically connected to its international – both European and Iberian – context.

Portugal's international status and the Portuguese foreign policy, following the French invasions and during the liberal revolutions, did not deviate from the strategic principles nor the diplomatic practice that historically had been defining.[1] At the level of strategic principles, the balance of power in the peninsula, and, inextricably linked to it, the balance between the continent and the Atlantic. And, consequently, the assessment of the alliances, among which the English Alliance continued to occupy a central place. At the level of diplomacy, Portugal, as a small power, continued to practice a diplomacy of opportunities, that is, to proceed according to conjunctural and correctable positions in a constant search for new answers to the new contours of the international conjuncture and the opportunities it offers.

During the last years of the 17th century and until 1807, Portugal managed to preserve the geopolitical balance between the continent and the Atlantic, that is, a degree of equidistance and neutrality between England and the continental powers. In other words, it managed not to choose between the English Alliance on which colonial integrity in Africa depended and the security of maritime routes to Brazil and the Franco-Spanish bloc on which peninsular sovereignty and integrity relied on.

Everything changes when, in 1806, Napoleon Bonaparte issues a decree proclaiming the Continental Blockade, a measure that involved closing all European ports to the British Armada, the confiscation of all English goods and the imprisonment of all British subjects. Portugal resisted, to no great avail. Finally, it complied with the closing of the ports, but not with the confiscation of goods or the imprisonment of British citizens. The bottom line was: war against Napoleon and the French invasions, on the one hand, and, on the other, England's ultimatum, to which, left with no alternatives, it eventually complied. Portugal was forced to make a choice. And it chose the English Alliance. In October of 1807, it signed a secret Treaty with England marking the end of parity and the beginning of English dominance. Not only political and diplomatic dominance, but also dominance over the military instrument itself.[2]

Now, it was in this context that the French invasions, the Court's departure to Brazil, the appointment of Beresford as Commander-in-Chief of the Portuguese Army and his influence over the Regency in Lisbon and the Court at Rio de Janeiro all took place. It was also in this context that King John VI signed the 1810 Luso-British trade agreement that opened Brazilian ports to English trade.

Once the war was over and Napoleon was defeated, in 1815, the Congress of Vienna confirmed peace and re-established order in the whole international system, based on two fundamental principles: the balance of power and dynastic legitimacy. The principle of balance referred to the external question and translated into the definition of an international order dominated, in the continent, by the "Holy Alliance" of the imperial powers that had defeated Napoleonic France

– Russia, Austria and Prussia – which would be joined by England forming a "Quadruple Alliance", great maritime power and true hegemony, undisputed until the seventies. The Congress, or rather, this small and aristocratic ensemble of great powers on its behalf, assigned itself the mission of ensuring international stability and security, whenever they deemed the European balance to be at stake. They outlined spheres of influence and allotted among them the tutelage of small and medium powers according to their own strategic interests. Naturally, Portugal was integrated in England's sphere of influence. The principle of legitimacy, in its turn, referred to the internal order and established the primacy of the dynastic principle over the principle of nationality. That is, it granted the great powers the right to interfere in the internal affairs of the small States within their sphere of influence. Naturally, against the nations and in defence of legitimism. Now, whereas the Holy Alliance took on and applied those two Vienna principles, the same did not happen with England. It favoured a narrow interpretation of the first principle and failed to apply the second. That is, it intervened whenever political balance and geography between the powers were at stake, but it did not meddle in domestic affairs nor did it seek to intrude in the political nature of the regimes. Its interference was reserved to those instances in which domestic change in a small power might affect the international order itself. This principle guided British foreign policy all throughout the period, traversing several governments, from Castlereagh to Canning and to Palmerston, and it surely was not absent from the establishment of Liberalism in Portugal. In accordance with this general principle and the terms of the English Alliance, England's conduct toward Portugal remained consistent: to safeguard the stability of borders, which meant, for Portugal, to safeguard its national independence; and not to intervene directly in political infighting, which allowed some leeway in terms of autonomy. England's vital interest was in controlling the bar of the Tagus, essential for the defence of Gibraltar. Gibraltar, in its turn, was essential for the strategic control of the entrance to the Mediterranean, which the great naval power could not relinquish without jeopardising its interests and influence in the Mediterranean and the Middle East. As such, England's vital interest did not concern the political nature of the Portuguese regime, unless a change in domestic politics meaning some kind of convergence with the French government might challenge British control over the Tagus.

In short, the Portuguese could govern themselves as they saw fit, as long as the British strategic interest was not imperilled.[3]

In keeping with the spirit of Vienna, the Congress of Verona of 1822 took charge of the situation in the Iberian Peninsula and decided for an intervention in Spain with the intent of bringing order to the "liberal unruliness", that is, to restore dynastic legitimacy. The mission was carried out by France, since it pertained to its sphere of influence. For the great powers, Portugal was not a concern: due to the difference in size and potential, once the situation in Spain was under control, the whole of the Iberian Peninsula was neutralised. Portugal would follow suit. And England would take care of it, as it did according to its understanding of the Vienna principles.[4]

The above does mean that Great Britain did not exercise their control over Portugal, which it did. First, after the French invasions and until 1820, in a direct and overt manner, by means of the 1807 Convention, the Trade Treaty of 1810 and the political and military tutelage over the Regency with the transfer of the Court to Brazil. Later on, in the terms of its understanding of the principles of the Congress of Vienna. In 1820, the liberal revolution broke up with the English tutelage as far as domestic politics was concerned. Externally, it kept Portugal in the British orbit, according to the system of international relations issued from Vienna. During the civil war, between 1820 and 1834, external pressure on the country multiplies: not only from England, but also from France and, especially, from Spain. However, once the war was over and Liberalism emerged victorious, English hegemony was again felt, this time indirectly through the Quadruple Alliance. Signed in London, in April of 1834, between William IV of England, Louis Philippe of France, Pedro IV of Portugal and Maria Christina of Bourbon of Spain, its objective was the stabilisation of the peninsula and the consolidation of the liberal regimes in the Iberian monarchies. During this period, however, Portugal essayed a number of attempts at counterbalancing, in the continent, the British naval hegemony, such as the trade agreement signed with Prussia in 1834 and the normalisation of relations with the Holy See, in 1848. Nevertheless, until 1851, the English Alliance continued to be practically the sole basis of Portuguese foreign policy.

The peninsular context weighed no less on the process of implementation of Liberalism than the international context. Right from the start, by the strategic unification of the Iberian Peninsula accomplished by the peninsular war and, further ahead, by the legacy it left of ideological influences and mutual political and military support between the two conflicting sides of the border. And despite the fact that liberal wars in Portugal and liberation wars, as they were known in Spain, were autonomous, along the process, on several occasions, Portuguese liberals supported the Spanish and the other way around. As, later on, the Spanish absolutists would support the Portuguese and vice versa. It was the so-called "peninsular political system", which operated until the victory of Liberalism and the consolidation of constitutional monarchies in the Iberian Peninsula.

Such was the international context within which the liberal revolution and the civil wars in Portugal unfolded. Considering its central role in the whole process, what was, after all, the state of the Portuguese Armed Forces?

Despite the modernising reforms launched by the Count of Lippe – the great strategist of the Prussian school, made Marshall – General of the Portuguese army in 1762 – in the beginning of the 19th century, the capacity and operability of the Portuguese Armed Forces was anything but outstanding. Apart from individual acts of military braveness, the performance of the Portuguese Armies in both the Roussillon Campaign, in 1793, under the command of Scottish Marshall Skellater, and in the War of the Oranges, in 1801, under the command of the Duke of Lafões, seem to confirm it. In 1806, General Gomes Freire de Andrade had devised an ambitious reform plan for the military inspired by the model of national army issued from the French Revolution: *Essay on the method for the organisation of the army in Portugal.* It was never implemented. And the French invasions only made matters worse. With the Court in Brazil and "King Junot"[5] in Portugal, the structures and the authority of the State began to collapse, peaking in the 1820 revolution, a process which only the construction of the liberal state managed to slowly and gradually revert. Recovery took decades to achieve. The Army was not spared by the disaggregation. In 1807, Junot and the Spanish Generals that escorted him, Taranco and Solano, ordered the publication of a set of decrees that resulted in all but the disarticulation and disarmament of the better part of the army and the totality of the militias. The higher military personnel had either left for Brazil with the Court or would depart compulsorily later on, to join the Portuguese Legion in combat alongside the Napoleonic armies.[6]

It took the arrival of the English to carry out the new military reorganisation. It bore the mark of the Duke of Wellington and its successor William Beresford, and signalled the dawn of a new cycle for the Portuguese Armed Forces. At the time, the basic structure of the army was still a heirloom of the Duke of Lafões, organised in three lines of defence: the first, comprised by the regular army according to the Count of Lippe's model; the second, by the militias – a new name, after all, for the old auxiliary *terços*; and, finally, the *ordenanças* (literally, ordinances), which were the territorial recruitment reserves.[7] What little remained of that Army after the invasions, however, was scattered and weakened, disheartened and unruly. That was the primary goal of the English reorganisation as early as 1808: to centralise and to discipline.

In the absence of radical changes or major reforms ("War time is no time for weapon cleaning", as the Portuguese saying goes), they sought, on the one hand, to reduce the power and influence of the old manorial and municipal structures on the recruitment and arrangement of troops and the arming of civilians, as well to strengthen the central administration of military institutions. On the other hand, they strove to reinstate hierarchy and military discipline, as well as to reinforce a single and solid chain of command At the same time, a latent tension was rising within the Army between the traditional criterion of the aristocratic social status – the nobleman as military leader – and a new criterion based on occupation – the military as professional career, which emerged following the end of the invasions and peaked during the liberal wars and the period of military intervention in politics.

With the expulsion of the French, once the war was over, Beresford's reforming initiatives did not halt. Quite to the contrary, a member of the Council of Regency and appointed President of the Council of War in 1811, he saw his powers enhanced and carried on his reforming work, centralising and training military forces. The task at hand was now to organise and place under a single command the numerous military corps of the regular army that had risen in a disperse manner in the various governing juntas, but also to control the irregular military, the militias and guerrillas that had fought against the French. In 1814, he issued a decree reshaping the military, in an attempt to adapt it to the needs of peacetime: 40,840 soldiers for line troops.[8] Only in 1816, however, did two regulations, one concerning the Army and the other pertaining to the *ordenanças*, foreshadowed a true military reorganisation in terms of a modern national-based army.

The recruitment system kept military service compulsory to all able men above a certain height, aged between 17 and 30 years. The old system of "casting lots" was also preserved, as well as perks and privileges, which abounded. As a last resource, there was always the possibility of paid exemption. Only those deprived of privileges, exemptions or money ever reached the ranks. For those who did, the length of service was unknown. The regulation failed to mention this aspect, giving rise to all sorts of interpretations and often to excesses and arbitrariness.

As to military organisation, the country's *ordenanças* were grouped into twenty four districts, and each district divided into eight captaincies-major and each captaincy into eight companies. Line troops were thus recruited from the *ordenanças*: from each of the twenty four districts, one infantry regiment, one cavalry regiment and one battalion of *caçadores* (hunters; elite light infantry soldiers). And from across the national territory, four artillery regiments. The recruitment of militias was region-based and from that contingent the police, in times of peace, and territorial defence, in times of war, were enlisted.[9] The regulation also stipulated the lodging of troops and the geographic distribution of military units: which regiments and battalions should be stationed in the recruitment district headquarters. This was a modern principle of territorial settlement and occupation whose purpose was not only external defence, but also, in a logic of reinforcement of the state apparatus, internal control of the territory and the populations.

The Navy was a world apart. The Council of the Admiralty, created in 1795 in line with the British model, was never fully functional in Portugal. With the Court's departure to Brazil, in 1807, it took responsibility over the affairs of the Navy. In truth, however, throughout Junot's stay, the French were the ones in charge, and after they left, the English took their place. In 1810, English Admiral Berkeley ruled the Council and, in 1812, was replaced by George Martin, who revealingly began to present himself as "commander admiral of the Portuguese Navy". The *Armada* possessed a cadre of officers from the Royal Navy Academy, and the remaining garrisons were composed by the Royal Navy Brigade and the sailors and cabin-boys sourced from military recruitment in the same terms as the Army and the reforms under way.[10]

Beresford's reforms were indeed forward-looking, not only insofar as they aimed at the professionalisation of the military career, but also because they envisaged the autonomy of the military institution. He replaced the traditional notion of class privilege with the modern logic of professional competence. Reactions did not take long to arrive: what meant "discipline", for the English, in the eyes of the more old-fashioned was "an offence against one's honour".[11] The feedback, however, was not limited to the traditional sectors. Beresford's despotic and insolent style, in addition to the haughtiness of the English officials, spread discontent among the Portuguese soldiers. Spurred by the growing institutional autonomy of the Army, promoted by Beresford himself, the corporate protests began to mingle with a nationalist reaction. One thing was becoming clear: it was time to "kick the English out". The first sign of revolt in the Portuguese ranks was General Gomes Freire de Andrade's conspiracy against Beresford, in 1817. The conspiracy was crushed and the General was savagely hanged in the fortress at São Julião da Barra. Far from curbing discontent, it further ignited the outrage. Many army men approached the liberal politicians and were welcomed at the Sinédrio (a secret association with the purpose of preparing a revolution). On the 24th of August, they were among the leading figures of the Revolution of 1820. Out of the dealings between civilians and army men who partook of the same ideals of Freedom was founded the Provisional Junta of the Supreme Government of the Kingdom, which, based in Porto, was destined to govern Portugal on behalf of the King. Symbolically, it was headed by an officer: Brigadier António da Silveira.

Revolution and counter-revolution: Absolutism, liberalism and civil war

The revolution of 1820 opened a new cycle in the history of the country: that of the liberal wars. Concomitantly, it also opened a new cycle in military history, marked by the prominence and intervention of the military in politics.[12] On the 15th of September, this time in Lisbon, a new military coup (*pronunciamento*), led by a junior officer, Aurélio José de Medeiros, which gathered the support of several military units and resulted in a second *junta*. In accordance with the Alcobaça agreement, in October of 1820, the north and south *juntas* were merged to form a single Provisional Junta of the Supreme Government of the Kingdom and a second Provisional Preparatory Junta of the Cortes, which were not supposed to be convened and integrated by the three States, but rather elected and constituent.

Soon dissidence emerged within the new liberal power. Almost from the start, between the more radical politicians and the more moderate army officials, gathered at the Rua de São Francisco Club. On the 11th of November of 1820, the *Martinhada*,[13] a new insurgency led by field Marshall Gaspar Teixeira de Magalhães Lacerda, overthrew the *Vintista* government and replaced it with a more moderate one. The Army, making use of its autonomy, appeared to be taking an institutional stand and enforcing its political will. That is, the military institution seemed to be asserting a political clarification in the heterogeneous ground of the 1820 victors. Days later, however, on the 17th of November, a countercoup took place: a new rebellion ousted the moderate government and gave the power back to the fallen government. The *Martinhada* had fallen short and the Army had not conducted itself as an institution after all. Beresford's reform had failed to completely unify the military institution: old factions and clienteles inherited from the manorial and municipal decentralisation carried out in the scope of the former regime's military organisation persisted still, and the tensions between the army nobleman and the career serviceman escalated. In the forthcoming liberal wars, not only the military were divided, siding with the different clashing parties, politicians also sought the military's support to affirm their legitimacy.

Despite this, the *Vintismo* managed to carry out its own military reforms: first, in recruitment, and secondly, in the very structure of the Army.[14] Recruitment was a burning issue for soldiers and junior officers in particular, but reflected on the whole of the troops' morale. The problem was that the law was silent regarding the length of military service, causing uncertainty, injustice and discontent. The matter, which had been dragging since Beresford's 1820 regulation, was so pressing that one of the first measures of the Cortes, in 1821, was to finally settle it. By the decree of the 17th of April, the Cortes defined with precision what should be the length of service: nine years for infantry and seven years for cavalry and artillery. Volunteers might enjoy a two-year reduction. The reorganisation of the Army came after and the new measures stipulated the dissolution of the old structure along three lines: the decree of the 16th of May suspended the recruitment of militias and disbanded the existing ones; and the decree of the 18th of August simply extinguished the *ordenanças*. The position of commander-in-chief in peacetime was also abolished. Later, in 1823, faced with the chance of an external invasion and the threat of internal resistance, the first-line troops were reinforced and national guards – which were supposed to mirror the liberal principles of the new power – were established. The law was one thing; the harsh reality was another, however. In truth, neither the national guards were implemented, nor the militias and the *ordenanças* completely disbanded. They continued to play a role in the recruitment system and came back strongly during *Miguelism*.

The *Vintist* reform also extended to the Navy: the Cortes dissolved the Admiralty Council and retrieved the old higher organisation of the *Armada*, assigning a Major-General with the duties of administrative management and military command. Recruitment maintained the general principles the decree had stipulated for the Army, but the Armada personnel, between officers and privates, was insufficient, numbering less than one thousand men: 925.[15] In this scenario, the sea was not a cause for concern, and so the Army was top priority.

The King had returned from Brazil in July of 1821, and sworn to the bases of the Constitution, as if conferring dynastic legitimacy to the liberal revolution. And, in September of 1822, the Cortes approved the Constitution. It stemmed from the will of the people and, in accordance with the *trias politica*, it granted primacy to the legislative branch and designated a single Chamber. In this manner, it deprived the King of the Chamber of Peers and denied him the right of veto over the legislation approved by the Parliament. The 1822 Constitution was "almost republican".

Reactions were not long in arriving. In the name of royal "dignity", Queen Carlota Joaquina refuse to swear to the Constitution, and the Cortes ordered her into exile. Miguel, her youngest son, shared her political views. Resistance surfaced within the Royal House itself, but also in the Church and, after the elections of the 22nd of November, in the Parliament. Resistance now counted on the members of Parliament who called themselves "realists" and whom the liberals refer to as "hunchbacks". Brazil's independence resulted in part of the Army joining the Resistance. In the early 1823, General Silveira led, in Trás-os-Montes, the first military movement against the *Vintista* government and in favour of the restoration of the royal power. The "regenerating army", as its own commander called it, comprised no more than one thousand men, between regular forces and scattered guerrillas, and was defeated by the constitutional forces led by General Luís do Rego and forced to retreat to the other side of the border. Military defeat, however, did not obliterate the political signal of the resistance's growing strength. The *Vilafrancada*, months later, turned the resistance into counter-revolution.

In accordance with the provisions of the Congress of Verona, France invaded Spain and, on the 23rd of May, barged in Madrid to put an end to the constitutional whims and restore the "legitimate authority" of Fernando VII. In Lisbon, the news caused panic within the government and got the realists moving. On the 27th of May, an infantry regiment rebelled and withdrew to Vila Franca. Prince Miguel accompanied the rebels, who saw in him their leader, and made a public announcement to the country. Subsequently, General Sepúlveda headed to Vila Franca with almost the entire Lisbon garrison. And, finally, the King, who thus politically confirmed the position of Prince Miguel. The government was dismissed and it was solemnly announced that the 1822 Constitution was suspended and was to sustain amendments. Prince Miguel was appointed Commander-in-Chief of the Army. It was his first great political victory. He kept this position until the *Abrilada* of 1824. As to the military, the *Vilafrancada* regime did not alter the 1821 decrees.[16] It did, however, reactivate the militias and *ordenanças* that had never actually ceased to exist, and dismantled the national guards that had never really affirmed themselves. Concerns were clearly political and Prince Miguel's supporters were gradually occupying key positions in the state apparatus. On the 29th of April of 1824, drawing on a *Vintista* conspiracy, they staged a coup d' état: the *Abrilada*. The military forces occupying Lisbon welcomed Prince Miguel and the persecution and arrest of liberals began. On this occasion, he had no support from the King or from the diplomatic corps. With the backing of the British, French and even the Spanish ambassadors, João VI took shelter in the English ship Windsor Castle, anchored in the Tagus. Safe from military threat, in the middle of the river and in English territory, the King summons Prince Miguel aboard, relieves him of his duties and regains control of the Army. Defeated, Prince Miguel departs for Vienna in Austria.

The radical liberalism of the *Vintismo* had been defeated in the *Vilafrancada*; the *Miguelist* absolutism in the *Abrilada*. João VI continued to rule, seeking to reconcile what gradually appeared to be more irreconcilable: the executive and the legislative powers; the King and the Cortes. And, above all, the two clashing political factions. With that goal in mind, he had promised a Constitutional Charter, but died on the 20th of March of 1826 without ever issuing it. Or, for that matter, settling the succession question.

The King's death imposed a political clarification concerning succession and the Constitution. Succession appeared to be a difficult hurdle. Prince Miguel had been defeated and was sent into exile. Prince Pedro, however, campaigned for the independence and was Emperor of Brazil. The Regency eventually proclaimed him king. He knew dual monarchy was unfeasible, and so he accepted the crown only provisionally, until he was able to come up with a political solution. On the 23rd of April of 1826, he issued a Constitutional Charter; on the 30th, he ordered elections to take place and finally, on the 2nd of May, he abdicated in favour of his daughter Maria, aged seven years old, who would marry her uncle Miguel when he, the following year, completed 25 years old. The arrangement seemed perfect: "The liberals would have a Constitution; the realists would have a King".[17]

As to the Constitution, the Constitutional Charter of 1826 was thus a result of the King's will, the King himself being, as well as the Cortes, a representative of the nation. The principle of the separation of powers into legislative, executive and judiciary was preserved, but the King was invested with a fourth power that worked as arbitrator of the other three: the moderation power. The executive power belonged to the King and the legislative power to the Parliament, composed of two chambers. But not only could the King appoint freely the Chamber of Peers, he could also dissolve the Chamber of Deputies and veto laws. Moreover, it looked like a sound constitutional and political compromise. As it turned out, it was not. Rather, it worked as an instrument of polarisation. The Charter earned the consensus of the liberals and rejection by the absolutists. The Army, or what was left of it after three interventions – the liberal revolution, the *Vilafrancada* and the *Abrilada* – did not escape the polarisation and consolidated the division between the two political fields. Marshall Saldanha, appointed Minister of War, led the Supporters of the Charter among the military. As early as 1826, the military rebellions against the Charter began, first in Trás-os-Montes, then in Alentejo, mobilising entire regiments. The fights took place between 1826 and 1827. The liberal side usually had the upper hand, but was never decidedly victorious.[18]

Lacking a military solution, Pedro IV pursued a political compromise, in a last attempt to prevent civil war, which, in truth, was already under way. The attempted negotiation consisted in a proposal addressed to his exiled brother: should Prince Miguel agree to swear to the Charter, he would be allowed to return to Portugal. Once back in Portugal, his marriage to his niece Maria would be negotiated and he would be appointed Regent of the Kingdom. In the spirit of the Congress of Vienna, Canning reached a settlement with Metternich and, pressured by the Austrian Chancellor, Miguel accepted the proposal. He swore to the Charter, negotiated his marriage to his niece and, on the 3rd of July of 1827, he was appointed Regent. Arriving in Lisbon on the 22nd of February of 1928, he once again swore to the Charter. He did not honour the commitment, though. In March, he dissolved the Parliament and, in May, he convened the Cortes. He did it in the old-fashioned way, however: clergy, nobility and the people. The Cortes proclaimed him absolute king in July of 1828. The return of absolutism pleased neither diplomacy nor the armed forces. The ambassadors, particularly those from European powers counting on Lisbon, suspended their diplomatic functions. Rebellions (*pronunciamentos*) within the Army recurred. Military units commanded by liberals, which were in great numbers and were stationed mainly in the north, revolted and created a new Government *Junta* in Porto. The *Miguelista* government had taken steps in the field of the military, concerned mainly with the political control of the armed forces: it had dissolved a number of military corps that were known to be liberal, such as the Volunteer Battalions of Commerce and the national guards; and had created many others otherwise inclined, such as the Royalist Volunteers. But the Army was divided and the liberals controlled a part of the units. And so the *Miguelistas* appealed to the Royalist Volunteers and mobilised the militias and the *ordenanças*, outnumbering the liberals in both men and commanders. In 1829, after the victory over the troops of the Junta of Porto, Miguel reduced the army to sixteen infantry regiments, four battalions of *caçadores*, keeping the four artillery regiments, the engineer battalion and shortening the contingent of the militias.[19] The *Armada* was still not a priority for the *Miguelista* government. After the independence of Brazil, it had lost strategic importance and there was a conviction that internal problems were solved in land and decisions rested with the Army. Moreover, unlike the Army, the Navy remained faithful to the Court and showed no liberal inclinations.[20] Liberal resistance, however, persisted in the Army: rebellions spread to the Algarve, Madeira and the Azores, although registering no great military successes. Political setbacks were what truly weighed on the liberals. Internal divisions between factions, lack of funding and with its political leader away in Brazil, since the *Belfastada* episode, in 1828,[21] the Porto *Junta* was dissolved and the liberal armies under the command of Bernardo Sá Nogueira withdrew to Galicia. Around five thousand army men, many high-ranking officers and hundreds of civilians – between doctors, judges and lawyers – took shelter in Spain.[22] The Spanish government was ruthless: they were disarmed and ordered to return to Portugal within 30 days. Many returned enjoying Miguel's amnesty, but not the military commanders or the civilian elites. Those left, first for England and then to the Terceira Island in Azores Archipelago, where from 1829 they established a sort of liberal State in exile within borders. Miguel was well aware of the symbolic and real threat posed by that bastion of liberalism in Terceira. For that reason, he sent a naval expedition to the Azores with the purpose of stifling the liberal audacity. On the contrary, the Royalist squadron was decimated in Terceira. Registering around one thousand casualties, between dead, wounded and prisoners, it was forced to beat a retreat. In April of 1831, an unexpected occurrence changed the course of events. Pedro I of Brazil decided to abdicate in favour of his son Pedro II and to return to Portugal to save the dynasty. With support from France and England, that is, from Talleyrand and Palmerston, he raised a two thousand-pound loan with which he set up a ground force gathering Portuguese exiles and hired foreign soldiers, and a naval force, through the purchase of two great ships equipped with artillery pieces.[23] In 1832, he took on the Regency and appointed a new government headed by Palmela, in the Terceira Island. The liberals now had all that they were lacking in 1828: political union, military forces, war-funding, and most important of all, the dynastic leader. From Terceira they left to conquer the Azores and, in the summer of 1832, from the Azores to conquer the mainland, "to free the homeland". Pedro envisioned his troops as a "liberation army" and, more than a military rebellion, he imagined a national uprising.[24] It never came to be.

Miguel and his troops awaited the liberal Navy in Lisbon. But Sartorius, the English Admiral commanding Pedro's squadron, headed north and thus "the Mindelo braves" – 5,500 army men and many civilians – were able to serenely disembark in the place that gave them their name, north of Porto, and take the city. Porto became a mainland Terceira. From there, Pedro launched his military reforms with a view to assuaging the discrepancy in the dimension of the forces, seeing that the *Miguelistas* were ten times their number. Miguel had reinforced his line Army: 24 infantry regiments, 12 cavalry regiments, four artillery regiments and 12 battalions of *caçadores*. On the whole, it amounted to around 25,000 men.[25] Expectably, Pedro's first measure was to once again dissolve the militias and the *ordenanças*. Afterwards, the battalions of Royalist volunteers. On the other hand, he reinforced the line troops, summoning those on sick leave, hiring foreign mercenaries and creating national battalions faithful to the liberal cause.[26] In January of 1833, the Royalist forces besieged Porto. There were 40,000 Royalists within the encirclement against 17,000 liberals in the city. The siege continued, as did the combats, but in a fairly even manner and with no clear victories. In the summer of 1833, the fate of the civil war took a turn. And, as very often had happened throughout the history of Portugal, the key was the command of the sea. The liberal Navy had been reorganised and considerably reinforced. Firstly, reorganised according to the model of the Royal Navy, although a number of positions were vacant. Sartorius had been appointed Commander-in-Chief and Major-General of the Portuguese *Armada* in very terms of the 1822 decree issued by the Cortes. Secondly, reinforced with five new ships and new artillery. Finally, the decisive change: Sartorius had been replaced by the distinguished Admiral Napier, who had accepted to command Dom Pedro's fleet and would act with great strategic autonomy.[27] The bulk of the Royalist troops being centred in the Porto siege, the squadron of the English admiral headed toward the Algarve. Off the São Vicente Cape, in an unprecedented and risky naval manoeuvre, Napier attacked, directly and through boarding, the Royalist fleet, spectacularly overpowering it. In two hours, Dom Miguel had lost his naval force. The Algarve being depleted, a military force commanded by the Duke of Terceira was able to move on to the north and march into Lisbon. It entered the capital without meeting with any resistance on the 24th of July. Soon, in order to come to the aid of their comrades in Lisbon, the Royalists abandoned the Porto siege. The liberals now dominated the two major cities and the sea. The final blow, however, came from inland, with the internationalisation of the conflict. Once again, the so-called "political system of the peninsula" was in place. With Fernando VII's death, the conflict between constitutionalists and *Carlists* was reignited. And when the liberal supporters of Dona Isabel

II took charge and expelled Dom Carlos from the country, Dom Miguel decided to welcome Dom Carlos in Portugal. The outcome: the Spanish government now overtly supported Dom Pedro IV. The Royalists were now cornered, in the peninsula, between land and sea. The Quadruple Alliance will eventually bring about their international isolation in Europe. Signed in London on the 22th of April of 1834 between England, France, Spain and Dom Pedro on behalf of Portugal, the alliance ensured the constitutional regimes in the Iberian Peninsula.

On the ground, the last battles were being fought: Saldanha in the Beira, Terceira in Trás-os-Montes and Sá Nogueira in the Algarve. But the end was near. On the 18th of February, Saldanha defeats the Royalists in Almoster, and on the 26th of May, Terceira crushes them once again in Asseiceira.[28] On this occasion, the military defeat has political consequences. On the 26th of May of 1834 the Évoramonte Convention was signed, thus substantiating Dom Miguel's capitulation and the end of absolutism. The civil war was over. Dom Miguel was sent into exile, under the Regency of Dom Pedro, Dona Maria II was crowned and the Constitutional Charter was restored. Liberalism was definitely established in Portugal.

Two liberalisms in conflict: Constitution, charter and civil war

Perhaps surprisingly, the establishment of Liberalism entailed neither the end of the military's intervention in politics nor the end of the divisions within the Armed Forces. The coups persisted and so did dissension. The gap between absolutism and liberalism was replaced by another one, between two different views on liberalism: the Constitution and the Charter. Outside and within the Army, the ideological and political breech now separated the supporters of the 1822 Constitution and those who championed the Constitutional Charter of 1826, *Vintistas* and chartists, radicals and moderate.

During the short length of his regency, Dom Pedro did not manage to secure political compromise nor ruling stability. He died shortly after the liberal victory, in September of 1834. Dona Maria II was proclaimed Queen at the age of fifteen. Soon the riots began, as well as pressure from the radicals, leading to the dismissal of Palmela's moderate government.

The Armed Forces were not spared the instability that plagued the government. The *Armada*, in which the great majority of officers had remained faithful to Dom Miguel, was subject to a clean-up process in its high-ranking personnel, thus giving an opportunity to younger and liberal officers. In 1835, a project for the reorganisation of the Navy was essayed and a "permanent naval force" was created, with short-spanned effects, however, due to budgetary constraints.[29] The big problem was the Army. The state of public finances caused delays in the payment of the allowances and the delays sparked discontent in the military institution. The last straw was when, in 1835, the following government, under the Quadruple Alliance, was forced to send an expeditionary division to Spain. The Spanish government had provided assistance to the liberals during the civil war and now it was time for the Portuguese government to aid the Spanish liberals in the *Carlist* wars. Duke of Terceira, who was Commander of the Army, authored the unpopular decision, and the protests were not slow in coming.

Radical rallies proliferated and, on the morning of the 10th of September of 1836, the National Guard battalions, commanded by Fernando Soares Caldeira, gathered in Rossio, in Lisbon, and proclaimed the Constitution of 1822. The political nature and the radical penchant of the Guards were well known and did not cause surprise. Unexpected, however, was the fact that the troops mobilised to scatter the Guards and quell the rebellion actually refused to fight and joined the insurgents. The Queen, humiliated, was forced to swear to the Constitution and to dismiss the government and appoint a new one, in which the heroes of the September revolution stood out:[30] Sá da Bandeira and Passos Manuel. The international atmosphere was favourable and also had an influence on the course of events. In France, the revolution of July of 1830, which had brought to power Louis Philippe of Orléans, had triumphed, and, in Spain, the *Carlist* wars traversed a period favourable to the liberals. Among the great powers that mattered, only England remained sceptic and observant regarding the *Setembrismo* (liberal faction supporting the September revolution). Not so much due to the radical nature of the regime, but to its closeness to France, which it feared might jeopardise British control over the Tagus and its naval hegemony in the Atlantic and the entrance to the Mediterranean Sea. For England, the *Setembristas* were a sort of "French party" and the English Ambassador encouraged the Queen to resist, offering her the protection of the British squadron anchored in the Tagus, off Belém. On the 3rd of November, Dona Maria II withdrew to the Belém Palace, supported by a number of members of the diplomatic corps and within sight of the English squadron, and dismissed the government with the purpose of restoring the Charter. In the following morning, however, the *Setembristas* curtailed the possibility of a military intervention by the English, mobilising the National Guards, which cut off communications between Lisbon and Belém, closing off the Palace and the Queen. The National Guards and the *Setembrista* commons marched into Belém. The English troops stationed in the Tagus disembarked in Alcântara to halt them. Passos Manuel and Sá da Bandeira were forced to intervene: the Portuguese returned to the barracks and the English to their ships. The political outcome was clear, though: the *Belémzada*[31] had failed and so had the purpose of restoring the Charter. The Queen Dona Maria was once more humiliated. The opposition was extreme: on one side, royal sovereignty and the Charter; on the other, popular sovereignty and the Constitution. A new *Setembrista* government was then formed and Constituent Cortes convened. A new Constitution was on its way.

The military rebellions (*pronunciamentos*) did not cease and neither did revolts against *Setembrismo*. They still raged in Alentejo and Algarve, opposing the Chartist military and the Royalist guerrillas. In 1837, Saldanha and Terceira spoke up in favour of the Charter and endeavoured to march into Lisbon to "rescue the Queen", who they said was being held hostage. But the revolt of the marshals[32] was a military and political fiasco, and the *Setembristas* consolidated their position. In April of 1838, the Constituent Cortes approved the new Constitution, to which the Queen was forced to swear.

The 1838 Constitution was an attempt at political compromise between radicals and the moderate, between the *Vintista* Constitution and the Constitutional Charter. It did not ensue from the King's will, but sprang from national sovereignty. It preserved the separation of powers, yet abolished the royal moderating power, that is, the King's power to arbitrate. The legislature was divided into two chambers, but the King was deprived of the Chamber of Peers. Still, the document provided a Chamber of Senators. The King presided over the executive branch, but more than that, he kept the power to dissolve the Cortes and an absolute veto over all legislation. It was a tough compromise, but still a compromise, which resulted in a short-term lull in conflict. With the 1838 elections, the Chartists re-entered the Parliament and, in 1839, the government. By the end of the year, they were back in the Armed Forces.

It was precisely during the *Setembrista* governments, in the first half of the 19th century, and under Sá da Bandeira's political orientation and military command that the most important military reform of the liberalism took place. A reform devised in the spirit of the more democratic liberal ideology inspired by the model of national army issued from the French revolution. Regarding recruitment, the decree of the 7th of December of 1836 stated from the start how many young men the military contingent should recruit: 8,700. These recruits were proportionately distributed by the administrative areas of the country: districts, municipalities and parishes. The method of recruitment was also specified. Voluntary enlistment was encouraged, but to fulfil the necessary number there was "casting of lots" among 18 to 25-year-old men. There were still exemptions, although less frequently, and money remissions. This system actually worked, but it never managed to gather the required 8,700 men, which led to changes in recruitment law in 1838 and 1840.[33] The decree of the 26th of November of 1836 also reorganised the Army's structure, establishing the partition of the country into ten military divisions, the ninth and the tenth corresponding to Madeira and the Azores. Afterwards, through a series of decrees published between January and May of 1837, the reform was finalised, taking on a global character and encompassing all the sections. Firstly, the composition and territorial distribution of the different forces: the infantry was composed of thirty independent battalions, each comprising eight companies. Before, there were twenty line battalions and ten battalions of *caçadores*. The cavalry now included eight regiments – four lancer and four "hunter" regiments; the artillery had once again four regiments, but counted on three batteries positioned in the Islands to secure the defence of the Atlantic Archipelagos. The *Setembrista* reform extended further: it restructured the health service and military education. The old Academies – Royal Navy and Royal Fortification, Artillery and Design were replaced by the Polytechnic School and the Army School, with the purpose of providing technical and military training to the future officers. The Military College's curricula were also revised and adapted to the preparation of cadets. For the soldiers, most of them illiterate, regimental schools providing elementary education were created. Finally, in the spirit of equality that was the flag of that more radical liberalism, and which should prevail in a national army, a regimental dress regulation prescribed that uniforms should be the same for everyone within each force. And, in order to put an end to the traditional arbitrariness that prevailed in rank progression, a general and standard policy was introduced. In the case of infantry and cavalry officers, antiquity was the criterion. Technical arms officers were subject to examinations. As far as weapons were concerned, no great progress was made.[34] With only a few minor alterations, this reform was in place until the reorganisation of 1849, which, truth be told, never really took effect. That is, it constituted the basis of the Portuguese Army until the Regeneration in 1851.

Despite its focus on professionalisation and the institutionalisation of the Armed Forces, this reform did entail an estrangement between the military and politics. Not only because politicians continued to summon the military, but also because military leaders themselves had become political leaders.

The fleeting trade-off after the 1838 Constitution intended to end the polarisation and to establish a political centre. The idea cherished by Rodrigo da Fonseca Magalhães was that there should be no longer Chartists or *Setembristas*, only "law abiders" (*ordeiros*).[35] The "order" did not last long. Rodrigo da Fonseca's government, in 1839, was too Chartist to settle in the political centre, and sparked a new *Setembrista* rebellion (*pronunciamento*), in 1840.

In June of 1841, the Aguiar ministry arrives in power. The young Minister of Justice had been a radical demagogue, then a firm "law abider", and was now more Chartist than the Charter itself. His name was António Bernardo da Costa Cabral. On the grounds of the electoral victory in January of 1842, the Chartists held a peaceful protest, in Porto, to demand the restoration of the Charter. Costa Cabral goes to Porto and, on the 27th of January, makes a proclamation and declares the Charter restored. He gained supporters and, on the 10th of February, already a minister in Terceira's new government, he proclaimed the restoration of the Charter in Lisbon. With the Charter in place, he ruled in a system of personal dictatorship during four years, between 1842 and 1846. He managed to conjure the hatred of every political force: Royalists, *Setembristas* and the Chartists themselves, disillusioned with his despotic ways. The negative coalition of such disparate oppositions once again appealed to the military. Military uprisings in the countryside were back, in Torres Novas and in Almeida. They were crushed. It did not mean the end of rebellion, but rather the onset of a new civil war. The spark to light the fuse was the so-called Revolt of Maria da Fonte, in April of 1846.[36]

In the northern region of Minho, rising against the introduction of a new tax, but mainly against the prohibition of burials inside churches decreed by Cabral, a popular insurrection of surprising violence took to the streets. As church bells sounded the alarm, the mobs invaded the cities of Minho and vandalised head offices and government departments. From Braga and Guimarães, the rebellion spread across the country and revolutionary *juntas* proliferated. The country was under sword and fire. Costal Cabral reacted as would be expected: by using force. Rights and liberties were suspended and the military were called upon. However, the chief of government, who was also Minister of War and, as such, Commander-in-Chief of the Army, declared he had no capacity to respond and that the situation at hand required a political, not a military solution. That is, the solution was the dismissal of the government. In other words, he chose to turn his back on Costa Cabral rather than waging a civil war.[37] And so, four years of *Cabralista* dictatorship were over, not due to popular revolt, but to lack of support from the Army. It was blatant proof that political force still relied on military force and, consequently, that to be a party leader one had to be a military leader. Costa Cabral was no such thing, and so he asked the Queen to be dismissed. The Queen proceeded to appoint a moderate government led by Palmela, but as she had become politically associated with Costa Cabral, she feared a *Setembrista* victory might have negative effects on the Crown. On the 6th of October, the Queen herself staged a coup and handed the government to Marshal Saldanha, who provided greater assurance. The following day, in a military parade at the Terreiro do Paço in Lisbon, the troops acclaimed the Charter. The Chartists had conquered Lisbon, but the *Setembristas* had control over Porto, where they established a Provisional Junta of the Supreme Government of the Kingdom, headed by General Conde das Antas. Sá da Bandeira joined the revolution. The country stood divided in half, politically and militarily. The Chartists ruled in Lisbon, the *Setembristas* ruled in Porto. Saldanha's troops controlled the south, those of Conde das Antas and Sá da Bandeira controlled the north. A new civil war had begun.

The civil war of Patuleia lasted from October of 1846 to May of 1847.[38] It lacked the larger scale of the first civil war between absolutists and liberals. It mobilised neither the same number of troops nor the same military and naval means. The Armies did not surpass five thousand men on either side. And, despite having spread to the whole territory, splitting the country in two, the theatre of operations was centred in Estremadura in central Portugal. The most relevant battle was fought in Torres Vedras, on the 22nd of December of 1846. It was not a decisive battle, nor did the war have one to speak of. What turned out to be decisive was still another unexpected factor: the intervention of Royalist guerrillas. Led by General MacDonnell, Dom Miguel's ex-military commander, the guerrillas entered the civil war faking an alliance with the Junta of Porto and the *Setembrista* troops presumably to wreak havoc and take advantage of it. The entrance of the Royalists on stage put the Spanish government in a state of alert and justified the request for external intervention. Spain appealed to the Quadruple Alliance, England backed the request and France followed the allied powers. Towards the end of May of 1847, in response to a formal request addressed by the Queen of Portugal, the Quadruple Alliance intervened. A Spanish division commanded by General Concha came by land, entered Portugal and, on the 3rd of June, took the city of Porto. The English fleet came by sea and also disembarked in the Foz do Douro. A few days later, the Spanish General, the English Admiral and the presidents of the Junta of Porto, Loulé and César Vasconcelos, signed the Convention of Gramido. The Junta was dissolved and its troops were disarmed and granted amnesty. The civil war was over.

Two years later, Costa Cabral returned to power, once again leveraged by the Queen. In December of 1849, the Army suffered a new reorganisation. Territorially, the country was partitioned into three military divisions: Lisbon, Évora and Porto, each holding eight subdivisions. In the Azores and in Madeira, autonomous commands were established. The distribution of the various arms was made in the following manner: the line army now counted on eighteen infantry regiments, nine battalions of *caçadores*, two lancer regiments, six mounted *caçadores*, three artillery regiments, one engineer battalion, a telegraph corps, a medical command and one general staff corps. The national guards remained in place, but were converted into second-line forces, a permanent territorial reserve force. However, the greatest novelty was the definition of the contingent according to different contexts: peacetime and wartime: 29,000 men in peacetime and 53,000 men in wartime. This was a visionary reform, both technically and organisationally, with the new structure it prescribed and the adjustments to incorporate technical innovations in engineering and communications. And also from the political military point of view, with its clear separation between peacetime and wartime. It was never fully effective, though, due to circumstantial political constraints.[39]

Cabral's second sojourn in power was fleeting. In April of 1851, Marshall Saldanha rebelled once again and, after a first failure, was forced to take shelter in Spain. Upon returning to Portugal, however, he entered Porto and was acclaimed by the insurgent troops. He then marched into Lisbon. The Army joined Saldanha. Days later he was summoned by the Queen to form the Ministry. The Regeneration had begun and, with it, a period of national reconciliation. Saldanha, appointed Commander-in-Chief, stated with the Army: he preserved the officers that had supported Cabral, but reinstated the *Setembrista* ones, promoting all of them to the next higher rank. The "monster-promotion" was the beginning of reconciliation in the Armed Forces. The Additional Act to the Charter, in July of 1852, was the beginning of reconciliation nationwide. The Regeneration was a period of peace, stability and development. It closed the cycle of military interventions and opened a new one marked by a "civilianisation" of the political order.[40] The "reign of phrase and gunshot" was over.[41]

CHAPTER
2
Building the African Empire

The new Brazil

The Independence of Brazil left an enormous void in what had been Portugal's imperial project. In 1825, when Portugal officially acknowledged Brazil and closed the Brazilian cycle, what remained of the Portuguese empire was a handful of small territories, scattered across the world, unrelated to one another and whose dealings with the metropole were next to none. The great majority of those territories were located in Africa and, apart from the Cape Verde and São Tomé e Príncipe Archipelagos, amounted to little more than a few villages or enclaves by the seaside, with the exception of two penetration zones in the interior of the continent: one in Angola, to the north of the Kwanza river, from Luanda to Malanje; and the other in Mozambique, in the Zambezi basin, between Quelimane and Tete. The colonial model remained that of the Old Regime: the predominant economic activity was the slave trade and political sovereignty was scarcely more than formal.

Apart from the eastern colonies, which had already enjoyed their heyday, Africa was all that remained to Portugal. It is no surprise, then, that not before long the African colonies were seen as a potential "new Brazil". The issue had previously been debated by the Cortes when discussing the Independence of Brazil, and the notion came up as a viable way to make up for it. During the twenties, a number of governments expressed their concern over the consolidation of Portuguese rule in Africa. With no great consequence, since the persistent atmosphere of civil war neither enabled the establishment of political consensus nor provided the economic means required by the colonial project. Only upon the end of the civil war and the implantation of Liberalism was it possible to attempt, for the first time after India and Brazil, a new imperial project, this time directed to the African continent.

The Marquis of Sá da Bandeira had a leading role in the whole process. The draft project was presented to the Cortes as early as

1836, and was imbibed in the legislation passed by the *Setembrista* regime between 1836 and 1838. The cornerstone of the new project was the abolishment of slavery. On moral grounds, naturally, as fitted Sá da Bandeira's progressive Liberalism, but also for economic reasons: the slave trade monopolised all the capital, annihilating all other economic activities, and thus thwarting the development of the colonies. The proposed economic model, to the contrary, relied on the agriculture of tropical products and the development of trade based in the overseas exportation of Portuguese products. Concomitantly, it purported to reorganize the political and administrative structure of overseas territories, in terms of both central and local powers. It was becoming clear, though, that none of this was feasible without controlling the territory.

By the decree of the 10th of December of 1836, the transatlantic slave trade was abolished. A previous decree, published on the 7th of December, had already reformulated the overseas administration. Finally, on the 17th of January of 1837, legislation was passed regulating trade relations between the metropole and the colonies. The promotion of the colonies was meant to bank on the creation of colonial companies and, especially, in white colonisation, which Sá da Bandeira sought to encourage as Minister of the Overseas, between 1838 and 1839. Such colonial policy implied a prerequisite that the minister also strived to encourage: the occupation of the territory, namely, the coastlines of the two main colonies – Angola and Mozambique.

However, Sá da Bandeira's plan and the three years of *Setembrismo* yielded few lasting results in territorial occupation and next to none as to economic development. Political instability and the crisis in public finances in the metropole, together with strong resistance from the colonies, led to failure. Internal political strife and lack of means deprived the central government of the authority and the instruments of the rule of law required to implement the plan. Local authorities took advantage of that incapacity, siding with the interests of slavers. The outcome was that the prohibition of slave trading was never really obeyed.

Not even the famous bill issued in 1839 by Lord Palmerston, British Secretary of Foreign Affairs, who claimed the right to seize the Portuguese ships involved in slave trading, managed to change the situation in any detectable way. Repression by the British fleet, unilateral between 1839 and 1842, and, posteriorly, in accordance with the terms of a British–Luso treaty, accomplished no more than diverting the routes. The trade itself persisted, as did the imperial model inherited from the Old Regime, throughout the entire length of the forties.[1]

The fifties changed this panorama, both domestically and internationally. Great Britain reinforced its expansionary policy and consolidated its trade network in Asia and South Africa, as well as the colonisation of Australia and India. France progressed in the Mediterranean and Northern Africa, investing particularly in the colonisation of Algeria. The abolishment of the slave trade and the fight for its eradication led both powers to intervene in the South Atlantic and to become closer to Sub-Saharan Africa. Interest in the African continent was no longer driven only by the humanitarian principles involved in the fight against slavery, but also by the budding economic opportunities.

Domestically, from 1851, the Regeneration closed the cycle of civil wars and launched a long period of political stability and "material improvements" that had an impact in imperial policy. Firstly, Portugal was fully aware of how European powers were drawing near the interest zones of Portuguese colonisation, and mindful of its own fragile position. Hence, all attention was directed to the slightest signal of colonial endeavour, by any of those powers, in the African theatre. Secondly, that concern became palpable and was materialised in the institution of the recreated Overseas Council, in 1851, and in a new colonial policy. The Marquis of Sá da Bandeira was yet again a leading figure. First, as the president of the Overseas Council, as early as 1851; and from 1856, as the Overseas and Navy Minister. The circumstances wherein the first colonial plan was devised in the thirties were different, but not its objectives. And the Overseas Council pursued them from the very beginning: first, to foster and strengthen the ties between the metropole and the colonies, by means of both market penetration and the employment of capital in the domain of production;[2] second, to promote territorial occupation and prevent, or at least minimize, the intrusion of the other European powers in Portuguese colonial territories, particularly in the zone between Angola and Mozambique.[3] These principles were materialised in a number of governmental measures. At once, the territorial occupation that aimed at seizing control of the Angola and Mozambique coastlines, and the penetration of a selection of inland areas. The Port of Ambriz, on the western coast, was taken in 1853. The objective was to, proceeding from there, occupy the Congo estuary and then the Bembe mines in the interior. Rather ineffectively, as it turned out, given British opposition, in the first instance, and indigenous resistance, in the second. In the eastern coast, the plan was to occupy the Arab trade ports on the north of Mozambique – Angoche, in 1862 – and attempt to go further inland, which also met with little success. Economically, the end of the Brazilian slave trade appeared to have removed the chief obstacle to a colonial economy based on agriculture, and to open up new possibilities. Measures were also taken to encourage trade relations between the metropole and the colonies. The creation of a shipping company – the Mercantile Union Company – was one such measure. Other measures intended to once more encourage white colonisation, which was granted a special fund. At the social and humanitarian level, the goals of abolishing the slave trade and the conscripted labour of "bearers" were retrieved. A decree issued on the 29th of April of 1858 went so far as to make a direct attack on slavery.

However, despite Sá da Bandeira's insistence on imperial policy, now as before, there were many hindrances and results were scarce. In the metropole, the ability to centralize and the State's authority were insufficient. And, in the colonies, the autonomy of local authorities and their connivance with slaving interests often led them to resist, if not outright disobey, the orders issued from the central government. Such was the case with the governor of Angola in relation to the decree of 1856 on the prohibition of the forced labour of "bearers". To these domestic obstacles, others of an international nature were added: the issue of the French ship *Charles et George*. Within the framework of the British anti-slavery policy and in strict compliance with the 1842 Luso-British Treaty, the governor of Mozambique order the capture, in 1858, of a French ship loaded with slaves which was headed for Reunion Island. The reaction of the French did not take long to arrive: they demanded the immediate release of the vessel and claimed for compensation, and, to dissipate any lingering doubt, they sent two warships to the bar of the Tagus. The government in Lisbon naturally expected support from England, which never came. Left with no options, the government gave in to the demands of the French. The internal consequences were immediate: the validation and reinforcement of the critics, who received confirmation of the double standard policy and the British hypocrisy in their humanitarian crusade against slave trade; and the delegitimisation and inevitable weakening of Sá da Bandeira's policy, which was grounded in the fight against slavery and cooperation with England.[4]

In early 1859, the government fell and, with it, Sá da Bandeira. The sixties would be marked by a profound change in colonial policy: not only was slave trade once more admissible, territorial occupation was also stalled. In the atmosphere of crisis that prevailed, there was even discussion on whether the colonies beyond the Cape should be sold as a way to capitalize the ones that were considered top priority: Angola and São Tomé. Things never reached that point, but by the end of the sixties, apart from the dominion over the Atlantic islands and the eastern possessions, little or nothing had changed in the effective occupation of the African continent. The same went for the imperial model. Real changes would only be accomplished in the seventies, again urged by a looming external threat: the scramble for Africa.

The scramble for Africa and effective occupation

The seventies were marked by a renewed interest in the African continent, an interest that was closely linked to the new circumstances that facilitated and encouraged colonial expansion at European level: demographic explosion, technological revolution, new economic and financial conditions and the new outline of the international scene.

The demographic expansion in the 1800s was a first factor. The swift decline in mortality and its subsequent stagnation contributed to a significant growth in the European population, which rose from 190 million in 1815, to 300 million in 1870. The colonial boom coincided with the demographic boom that supplied the great European migrations. And although the migratory surge was directed mainly to the Americas, it also impacted Africa.[5]

The revolution in technology was another decisive factor. In the first place, in the domain of maritime transportation, with the replacement of sail with steam, which at once allowed for an extraordinary increase in cargo capacity and shortening of shipping times and a reduction in service costs. And thus appeared the great shipping companies connected to the big industrial and financial interests.[6] Secondly, in the field of telecommunications, the advancements were also relevant and improved communications. Thirdly, colonial medical care and, in particular, the malaria prophylaxis, with the use of quinin, made it possible to overcome the threat posed by tropical diseases. Finally, innovations in weaponry and in military industries resulted in exponential increase in speediness, mobility and fire power for the European armies, granting an overwhelming military superiority. Technology made the number of soldiers practically irrelevant. So much so that it ensured military superiority even when discrepancy in numbers were 1 to 20. This is what made possible the colonial expansion with relatively small European contingents.

All this made it possible to overcome the obstacles that stood in the way of colonisation, from distance to climate, from topography to disease and the resistance of the native peoples. Other factors, namely economic, financial and relating to international politics, were likewise important.

The new international financial conditions also favoured colonial expansion, with the concentration of great amounts of capital that formed the basis of colonial investments. Their surplus balances led the great powers to export capital and these exportations took the shape precisely of colonial investments. From then on, whether deriving from private funding or public loans, European capital invaded the world economy, with Great Britain in the lead, followed by France and Germany.[7]

Finally, the international scene underwent a change that triggered colonial rivalries that fuelled the scramble for Africa. At the peak of European nationalisms, and the Italian and German unifications being concluded, from 1871 the international scene acquired a new configuration: Germany emerged as a power and British hegemony gave way to an international system tendentially multipolar, in which European rivalries reflected in the colonies.

Conditions were surely gathered for colonial expansion. But why did the Europeans decide to venture into such an enterprise? Why now, the scramble for Africa?

Economic, political and ideological reasons, to which was added the galvanising action of a series of missionary movements, were at the origin of the scramble for Africa.

Economically, the year of 1873 marked the opening of a new Kondratieff cycle, in a downward phase of the wave,[8] and inaugurated a period of recession for the developed European economies. Such a scenario precipitated the return to protectionist customs policies to the detriment of free trade, and urged those economies to conquer new – naturally extra-European – markets. Almost simultaneously, those countries entered a second industrial revolution, whose accelerated pace of development resulted in a scarcity, in Europe, of raw materials, all of them originating in the African continent. To the search for raw-materials and the conquest of new markets, another factor was added. The big investments jeopardised the great amounts of capital at stake, requiring an economic programming of resources that implied not only a geographic and geological acquaintance with the lands, but, more importantly, the effective occupation of the territories.

Politically, the reputation of States, the national "pride" of European powers and strategic reasons – conquering naval points of support to provide security for maritime routes and territorial domains – were among the most relevant causes.[9] More important still, the new international situation, following the unification of Germany, favoured the colonial expansion. In fact, after the Franco-Prussian War, in 1871, territorial expansion was completely blocked in Europe. Under the joint pressure of the nationalities movement and the so-called "system of compensations", the most timid attempt at expansion in Europe risked degenerating into a serious conflict. The idea then came up that the interplay of the European "balance of powers" might be extended to extra-European zones, zones known to Western legal experts as *res nullius*. From that moment, the "European balance began to be played outside Europe".[10] And, in that game of extra-European territorial expansion, the African continent began to play a key role.

Ideologically, encouraging and validating the colonial enterprise, the great colonial ideologies and imperial projects began to be forged. These projects were designed around key points and geopolitical lines such as the Cape to Cairo imperial connections, in the British case, Dakar to Djibouti, in the French case, or the Portuguese Meridional Africa. Often taking on, ideologically, the transcendent significance of a "historical mission" or a "national destiny".[11]

All of it resulted in a major change in the imperial model: the Old Regime imperialism, informal and based on market penetration, gave way to a modern, formal imperialism that required territorial occupation. In light of it, the onrush of the European powers towards Africa, in the seventies and eighties, is unsurprising.

The first European expeditions, in this new cycle of African history, were led by the Scottish missionary Livingstone who, between 1840 and 1873, conducted a reconnaissance of the entire course of the Zambezi River, the Malawi and the Tanganyika Lakes up to the source of the Zaire River. And also by Stanley, a Welsh journalist based in New York, who, after a first trip with the purpose of meeting Livingstone, which he did, established that the Zaire River was navigable from the Tanganyika Lake to the western coast, thus opening up prospects for European penetration in the African continent. Another worth-mentioning explorer, the Italian-born French citizen Brazza, carried out an expedition in 1875, funded by his government, to the Equatorial zone north of the Zaire River.

However, before the great European powers began to carve up the continent, history had already registered twelve Portuguese expeditions into the hinterland between Angola and Mozambique, from 1785 to the great explorations of Capelo and Ivens, in 1877. Among these, a few stand out, such as the one led by Francisco José de Lacerda e Almeida, from Tete to Zimbabwe, in 1797, Pedro João Baptista and Amaro José's, from Caçanje in Angola, to Tete in Mozambique, between 1802 and 1806, Correia Monteiro and Pedro Gamito's, from Tete to Cazembe, in 1831, and the crossing from Bié to the opposite coast by Silva Porto, in 1853.[12]

The Portuguese pioneerism in Africa, in terms of international law, translated into the so-called "principle of historic rights", which, being internationally acknowledged, regulated a number of colonial matters until 1875. It was indeed under the principle of historic rights that diplomatic solutions were sought to settle the first colonial Anglo-Portuguese disputes, which were also the first challenges to Portuguese sovereignty in Africa. The first one – the question of Ambriz – was solved by the Portuguese government by means of a military expedition that resulted in the effective occupation of the territory in 1853. The other two, however, required international arbitration: the dispute over Bolama Bay was sent to President Grant of the United States of America who, in accordance with the principle of historic rights, ruled in Portugal's favour on the 21st of April of 1870; and the dispute over the Lourenço Marques Bay, decided by General MacMahon, the French President who, according to same principle, acknowledged Portuguese sovereignty over the disputed territories on the 24th of July of 1875.

The principle of historic rights was therefore the principle of international law that regulated colonial disputes until 1875. In 1883, Luciano Cordeiro defined the concept in terms of "discovery/ownership/recognition" and concluded that "these have been the principles invoked and sanctioned by the States that contemporarily have sustained and settled international disputes, and these are the titles in which the sovereignty of cultured nations is founded and exerted over various inhabited territories, populated by savages or colonised by those nations".[13]

All that changed from 1875. In 1876, at the Brussels Conference, the principle of historic rights was challenged for the first time, and in the decade that followed and until 1885, a new principle of international law emerged regarding colonial issues: the principle of effective occupation. It was inherently linked to the colonial scramble by the European powers, which, moreover, it aimed to sanction internationally.

In fact, from 1875 and, vertiginously, between 1880 and 1885, the Europeans engaged in a race for Africa. From 1881, France penetrated further into the interior of the continent and occupied the vast basin of the Zaire River. In northern Africa, besides Algeria, it settled in Tunisia. From 1882, Great Britain settled in Egypt and, from South Africa, initiated a project to cross Africa from south to north that would culminate in Cecil Rhodes's Cape to Cairo plan. Belgium, through the International Association of the Congo, began to occupy its place soon after the Brussels Conference, in 1876. Germany only converted to the colonial imperative from 1884.[14]

The race for Africa was under way: the "Scramble for Africa", released and popularised by the newspaper *The Times*,[15] which had a

French version in "*Course au Clocher*".[16] It was clear that modern imperialism required territorial occupation.

The change was notorious, and Salisbury himself acknowledged it in 1891: "When I left the Foreign Office in 1880", he remarked, "Nobody thought much about Africa. When I returned in 1885, the nations of Europe were almost quarrelling with each other as to the various portions of Africa they could obtain."[17]

All this movement had profound repercussions for Portugal and aggravated the already grave sense of fragility raised by Portugal's position in Africa. To begin with, Portugal had been left out when Leopold II of Belgium convened the first Brussels Geographic Conference. It seemed preposterous, for a country that ever since the independence of Brazil had construed a colonial imaginary centred in Africa. On the other hand, when Livingstone and the other European explorers publicised the reports of their expeditions and claimed to be pioneers in their endeavours, Portugal took it as a humiliation. Not only for overtly challenging the legitimacy of the principle of historic rights that it claimed for itself, but also for baring the state in which many Portuguese territories in Africa could be found.

The spectre of the loss of the empire was materialising and, with it, the "spoliation myth".[18] It is under the shadow of this myth that Portugal joins the land-grabbing race in Africa. Ten years later than its European counterparts, the Lisbon Geographic Society was created in 1875 and, soon after, the Central Permanent Commission of Geography. The former was a private association, the latter depended on the Overseas and Navy Ministry, yet the two were eventually merged, four years later, under the sponsorship of the Minister Andrade Corvo, the new champion for the Portuguese imperial project. The Geographic Society became, from then on, the veritable colonial lobby in Portugal.

The first great Portuguese expeditionary journeys to the interior of the African continent began in 1876, under the auspices of the Geographic Society and its lifelong secretary Luciano Cordeiro. These were known as "public work expeditions": to survey, to chart, and then to occupy the territory, such were the guidelines. In 1877, Serpa Pinto, Hermenegildo Capelo and Roberto Ivens left Benguela and headed for Bié, where they parted, Capelo and Ivens going north and Serpa Pinto going south onto Durban. The Portuguese expeditions went on until the Berlin Conference: Henrique de Carvalho explores the whole of the Luanda region in 1884, between 1884 and 1885 Capelo and Ivens travel from the western coast to Tete, and Serpa Pinto and Augusto Cardoso undertake an extensive exploration of the Nyassa, Shire and Zambezi areas down to Quelimane.[19] Back in the metropole, these explorers are welcomed as heroes. A crowd cheering them upon their arrival, and an agenda filled with audiences with the King, public conferences and official dinner parties. Afterwards, travels to European capitals, to publicise the scientific findings.[20] Many of the scientific objectives were not achieved, but that mattered little. The important thing was the heroic memory they had left and the new impetus they had imparted to the Portuguese imperial project. An impetus that, once more under the tutelage of the Geographic Society, met with a new plan for territorial expansion in 1881, titled "Appeal to the Portuguese People for the Sake of the Honour, Law, Interest and Future of Portugal". The pretensions of the "Appeal" were in line with the grandiloquence of the title. The territorial claim encompassed no less than the whole left bank of the Zaire including the mouth of the river, in the western coast; in the eastern coast, control of the Nyassa as a way to expand the sphere of influence and the penetration into the interior. It yielded no immediate results, but illustrated clearly the new Portuguese colonial ambition. Andrade Corvo, the Overseas and Navy Minister, had a strategic vision for the Empire that consisted after all in an upgrading of Sá da Bandeira's ideal seasoned by political realism. The territorial expansion had to take place; however, not through the use of military force, but rather by the influence of civilisation. Not by conquering, but rather by integrating the indigenous peoples.[21] Corvo was aware of both the international conjuncture and the country's limited resources. For that reason, he knew that the success of Portuguese colonial policy relied on openness to the world, to trade and to foreign investment, crucial for the development of the colonies. Still, Andrade Corvo was also the Minister of Foreign Affairs, and so he sought that openness in the scope of international cooperation. Conflicts with England over the slave trade had been overcome, and protectionist restrictions on customs had been revoked. Therefore, there was nothing preventing the revitalisation of the "old alliance" or keeping colonial openness from being wielded in that framework. And so did Andrade Corvo under the aegis of a series of treaties signed from 1878. The first, concerning India, envisaged the connection of the Mormugao Port to the British railway line, with the purpose of stimulating economic relations between the Portuguese colony and the Indian hinterland. The second, in 1879, related to the funding of the construction of the Port of Lourenço Marques and the railway line to Transvaal, in exchange for commercial benefits in the Zambezi River and customs exemptions for British goods at the Port of Lourenço Marques. And, lastly, a third treaty, signed in 1884, which attempted to define Angola's northern border. Great Britain, pressed by the Franco-Belgian advances in the region, finally conceded the

Portuguese claims over the left bank of the Zaire River and acknowledged Portuguese sovereignty over "the western African coast between 5°12' and 8° of south latitude, extending to the interior of the Zaire River onto Noqui and, from there, to the limits of the possessions of coastal and marginal tribes".[22] That is, between the Ambriz in Angola and Porto da Lenha in Congo. In return, British goods would enjoy ten years of customs exemptions.[23]

Andrade Corvo's policy met with no success. The agreement concerning India did not affect the imperial projects of the European powers, and, in any case, was not central to Portuguese colonial policy, but the other two were complete fiascos. The Lourenço Marques Treaty, under the crossfire of the oppositions that converged in imperial nationalism, failed to be ratified by the Parliament, in 1881. The treaty of Congo, in 1884, caught between internal opposition and diplomatic contestation by France and Germany, newly awakened to the colonial imperative, fell without even being submitted by the government to discussion in the Parliament.[24] Surprised by the British stand, but forced to accept it, the new Portuguese Minister of Foreign Affairs, Barbosa du Bocage, declared that, at that point, only in the context of an international conference could a solution to the problem be found. In truth, the Berlin Conference emerged from international contestation against the Congo treaty and with the purpose of finding a solution for the questions it had left unanswered.[25]

In July of 1884, Chancellor Bismark, in line with the French government of Jules Ferry, took the initiative in convening, for November of the same year, an international conference in Berlin with a two-fold purpose: first, to regulate colonial trade, particularly that which relied on inland routes of fluvial penetration, such as the Niger and the Congo; second, to delineate the general principle of international law concerning the colonisation of overseas territories. The outcome of the Berlin Conference[26] opened a new page in colonial history and impacted the future of the Portuguese imperial project.

In the first place, and on the side lines of the conference, although enjoying the clause of "most favoured nation", the territorial sovereignty of the International Association of the Congo was recognised (14th of February) over the region that days later (23rd of February) would become the Congo Free State and whose sovereign would be Leopold II of Belgium. Portugal kept the territories on the left bank of the river, but was forced to relinquish all of lower Congo, which had been for a long time one of Portugal's claims on the basis of historic rights.

Even more important was the resolution that approved no more and no less than the international recognition of the principle of effective occupation, which, from then on, regulated colonial matters.

It bore heavy consequences for Portugal. One of them was the adjustments made to the borders of their possessions and those of the new colonial powers. These adjustments were enforced soon after, in 1886, through the signing of two conventions – one between France and Portugal and another between Germany and Portugal. Boundary agreements, these conventions did not exhibit signs of political or diplomatic approximation, nor did they betray the English alliance. They did not exclude the possibility, though, especially in the case of Germany. The second and graver consequence was the new colonial law issued from Berlin, enshrining the principle of effective occupation and awarding mere adjutory value to the principle of historic rights.

However, despite British pressure for the principle to be applicable to the whole continent, the negative reaction of all other powers prevented it, and so the final statement was restricted to coastal regions. Hence the imperative of expanding effective occupation to the hinterlands through agreements on "spheres of influence".[27]

As a result, Portugal was forced to effectively occupy the territories it claimed to own by historic right, that is, to establish its sphere of influence in the African continent. The project involved connecting the two coasts, from Angola to Mozambique. It was given the imperial name of Portuguese Meridional Africa and in its graphic representation Portugal's sphere of influence featured the colour pink, which eventually rendered it famous as the Pink Map.

From that time onwards, invariably under the auspices of the Geographic Society and the encouragement of the new Minister of Foreign Affairs – Barros Gomes – the project of Portuguese Meridional Africa was launched and new colonial expeditions were set in motion in order to implement the Pink Map.

From 1887, Paiva de Andrade occupied Zambezi, Vítor Cordon and António Maria Cardoso settled in Nyassa and other regions of north Mozambique, while in the western coast, Artur Paiva and Paiva Couceiro occupied Bié. Serpa Pinto, after occupying Tungue, undertakes his mission in the region of the Macololos, who he manages to subdue. This mission turned out to be the cause of a new conflict between England and Portugal.

In fact, the Conventions of 1886 had only served to negotiate spheres of influence with France and Germany. The same did not happen with Great Britain and, for that reason, the risk of conflict was a tangible threat, all the more so since the horizontal project of the Portuguese Meridional Africa impinged directly on Cecil Rhodes's vertical project of a Cape to Cairo connection, supported by Great Britain.

There had been changes also in terms of foreign policy. Taking advantage of the German emergence in the European scene as an alternative to English hegemony, Barros Gomes, the new Minister of Foreign Affairs, essayed a diplomatic approach to Germany. Not with a view to altering the English alliance, but to obtain negotiating leeway with regard to Great Britain.

The coincidence of the diplomatic approach to Germany, in Europe, and the expeditions to establish territorial occupation, in Africa, led to a rising tension between Portugal and England, which ended up degenerating into conflict.[28]

Between 1887 and 1889, England launched a diplomatic offensive with the purpose of moving out of the way the Portuguese obstacle that hampered the implementation of their project.[29] Portugal's diplomatic strategy was to defer the matter for as long as possible. It managed to put it off for two years. Having been constantly postponed, reaching agreement was no longer possible. In the African theatre, incidents recurred.

When Serpa Pinto arrived in the Shire Valley, he saw that an English flag had been hoisted and that the Macololos had rebelled against Portugal. Military operations went on and João de Azevedo Coutinho conquered the region and subdued the African chiefs, who yielded to Portuguese sovereignty. It was all it took for the British to regard it as a *casus belli*. They rejected all and every settlement or arbitration agreement, concentrated their naval forces in strategic points of the African coast and, on the 11th of January of 1890, they sent Portugal a memorandum demanding Portugal's immediate retreat from the disputed zone. It was the Ultimatum.[30] The Ultimatum and the disparity in strategic potential between the two powers left Portugal without an immediate alternative. The government had to give in to the English demands. To mitigate defeat, Portugal requested the international mediation or arbitration that article 12 of the General Act of the Berlin Conference conceded. Two preliminary requirements had to be met, however: first, Great Britain had to accept it; second, Germany had to agree to act as mediator. Neither one did so.[31] Left with no choice, Portugal was forced to negotiate directly with England. After a year and a half of negotiations, a botched treaty (20th of August[32]) and a *modus vivendi* (14th of November[33]), an agreement was finally reached accommodating the English territorial claims and defining the limit of Mozambique's western border. The Treaty of the 11th of June of 1891[34] put an end to the conflict. And, at the same time, to the Pink Map dream.

The crisis of the Ultimatum did not bring about changes to foreign policy, but caused a tremendous internal commotion. According to the national imaginary, the "spoliation myth" had become real. The country was devastated by a collective sense of humiliation and there were impassioned reactions against the "shameful outrage".[35] A national trauma that had two immediate consequences: the reinforcement of imperial nationalism, which held sway over a whole generation, and the urgency of colonial expeditions to occupy territory, now in the shape of military campaigns.

The metropolitan army and the colonial army

What was, then, the military instrument employed by Portugal in its so-called pacification campaigns? Ever since 1851, the Regeneration had brought the military back to their headquarters. The cycle of military interventions was over and a new one marked by a "civilianisation" of the political order began. With the "pacification" of the Armed Forces and their subordination to constitutional power, the ensuing military reforms aimed essentially at adapting the structure and the weaponry to the strategic and technologic evolution in the second half of the 19th century. From 1861, Fontes Pereira de Melo initiated a process of modernisation of the army's weapons, replacing existing ones with English gear. The old percussion lock weapons were replaced by Enfield rifles and carbines whose numbers Fontes added to with a second large order in 1866: 10,000 carbines, 2,000 with bayonet and 2,500,000 cartridges. On the other hand, since 1863 the artillery had been equipped with striated cannons, howitzers and mountain guns sourced from the army arsenal. As to the reorganisation of the organic structure, in 1869 a new reform was undertaken in line with the Prussian strategic principles rendered famous following the victory in the Battle of Sadowa. The military contingent was fixed at 30,000 men in peacetime and 50,000 men in wartime. The system of recruitment suffered no significant changes. Military service remained unpaid and compulsory for a period of five years. Upon its completion, the recruit was transferred to the militias and, from the age of 50, to the military reserve force. The "casting of lots" dictated who, among the drafted, would add to the required number. However, the system of money remissions remained in place, which meant that only the destitute actually engrossed the ranks. Its revenue was supposed to be allocated to the modernisation of weaponry. The organic of the army partitioned the territory into three military divisions, of which three were first class: Lisbon; Porto and Estremoz.[36]

In 1884, Fontes undertook a new reform that increased the headcount and set career progression for officers and graduates. It divided mainland Portugal into four military regions – Lisbon; Porto; Viseu and Évora – and two commands for the Archipelagos of the Azores and Madeira. The big change, however, came in 1885, with a new technologic leap in the field of armament. During the eighties a new technology spread across the European armies, which consisted in breech-loading weapons, particularly the Snider breech, which enabled much faster and easier loading of ammunition. Besides, another feature of the new Snider technology was that it was adaptable to the Enfield rifle, also made in England and employed by the Portuguese army. Fontes Pereira de Melo was able to purchase from the English Ministry of War 25,000 Snider-Enfield rifles plus 20,000 breeches to use with the Enfield rifles already in the Portuguese army arsenal. Artillery weapons were also updated. Between 1874 and 1875, twelve batteries (six Kreiner and six Krupp) had been purchased from Prussia, and in 1884, ten six-gun complete batteries, state-of-the-art technology at the time.[37]

With the Navy, things were not much different. The second half of the century is marked by advancements in naval technology, with the introduction of the metal-plated hulls and steam power, and the Portuguese Navy was forced to keep pace with it. In 1866, a new system of naval recruitment was established and, in 1868, the services of the Overseas and Navy Ministry were restructured. The new top priority being colonial policy, from 1875, the government, in an initiative by Andrade Corvo and Fontes Pereira de Melo, launched an acquisition programme that brought a measure of modernisation to the Portuguese Navy. More up-to-date means were purchased from England and even the navy arsenal: seven gunboats, two corvettes, one battleship and two torpedo boats. Some years later, in 1883, under the leadership of Pinheiro Chagas, the structure of the navy was reorganised, with the institution of two naval divisions: the Atlantic and Indian Ocean divisions. Aware that the *Armada* was in need of reinforcement, and Africa being at the forefront of concerns, the government commissions three more gunboats, three more torpedo boats and other minor ships. This upgraded naval force would play a significant role, not only as regards transportation between the metropole and the colonies, but also in military operations taking place overseas.[38]

Peace in Europe and the absence of any external threat meant that the military capacity of the Portuguese army had never been put to the test. This must leave us under no illusion as to its actual state. There was a widespread and consensual impression, in Portuguese society, that the army was in no condition to ensure national defence.[39] Only in Africa, in the course of the "pacification campaigns", had its capacity been indirectly tested. Indirectly, because in fact, one thing was the metropolitan army, and quite a different one the colonial army. Institutionally, they were separate entities: they had different structures and hierarchies, autonomous recruitment and training, and different equipment and armament. And only the priority given to the Empire and the African campaigns, from the eighties onwards, gradually changed this scenario.[40]

At the colonies, namely Angola and Mozambique, colonial troops were divided into capital's troops and provincial governments' troops. The capital's troops were composed of three battalions: one battalion of *caçadores*, one infantry battalion and one artillery battalion, consisting almost solely of convict exiles (*degredados*). The troops of the regional governments comprised eight companies composed of natives, for the most part forcibly recruited, and exiles from the kingdom. They lacked basic training and military discipline and therefore were useless in combat. Middle ranks, the sergeants, were recruited among the *mestiços* (mixed race) or the exiles from the kingdom, and even officers did not possess the same education and military training as the metropolitan army.

This military structure persisted until the sixties and only then, with the arrival of contingents from the metropole for the occupation campaigns, did it begin to slowly change. In 1869, the Mozambique Organic Charter is drawn up, and likewise with Angola. These charters established a new basis for military organisation resembling the one at the metropole: first-line troops divided into five battalions composed of indigenous soldiers and officers from the metropole; and second-line troops, composed of sepoys, land hunters and irregular troops lent by allied or subjugated tribal chiefs.

Based on the Organic Charter, and after the definite demarcation of borders in 1891, the military structure was reorganised in 1893 and completed in 1895, when the military campaigns were already in full force. The first-line troops were arranged in three battalions, with four companies of indigenous soldiers and one metropolitan artillery battery. The cadre of officers was also reorganized and opened to metropolitan officers, who might be commissioned for service. Recruitment also undergoes changes: from then on, all Portuguese males, between 20 and 30, resident in the colonies, were obliged to do military service, although money remissions – when affordable – were always an option. Second-line troops were recruited among the natives. The chiefs of their tribes were required, in wartime, to deploy and send them to the Portuguese authorities.

With regard to weaponry, as operations in Africa progressed and it became necessary to impose military superiority, colonial troops were provided modern weapons, which was decisive for the outcome of the campaigns. From the seventies, and especially during the eighties and nineties, Angola and Mozambique saw the arrival of new rifles and carbines, machine guns and the artillery.

The colonial army was reshaped one last time in 1901. Undertaken by the Minister Teixeira de Sousa, the reform established at long last, the institutional articulation between the metropolitan and the colonial armies. First-line troops in each colony now counted on a permanent corps from the metropolitan army, which, from that moment, would constitute the central core of colonial forces, militarily decisive in the victory of "pacification".[41]

This reform, long overdue, was unwelcomed by the metropolitan officers, who had put up corporate resistance to the establishment of a permanent military force in the colonies. With a two-fold outcome: not only did it make it necessary to resort *en masse* to the natives,[42] it also forced the deployment of consecutive military expeditions from the metropole to the most crucial campaigns. Probably at greater financial cost and with less effectiveness than if there had been a permanent force available.

The metropolitan military elite that excelled in the African campaigns would, moreover, draw out in the shape of a new avatar the tension between the European and the colonial military. Between the two components of the army: the bureaucratic component, featuring the military behind their desks in Lisbon, and the operational component of the Africanist military, turned into heroes by virtue of their colonial exploits.[43] Mouzinho de Albuquerque Ayres d'Ornelas and Paiva Couceiro are merely the more striking examples.

Beyond all the human effort and financial investment by successive governments, nothing translated so sharply the national priority given to the overseas empire and the "pacification" campaigns as the symbolic prestige and the political significance that the Africanist military earned among Portuguese society.

The "pacification" campaigns

The Treaty of the 11th of June of 1891 had defined the borders of Portuguese Africa, which remained practically intact until decolonisation in 1975. In the last decade of the 19th century, the Portuguese colonial empire comprised five possessions in Africa: the Archipelagos of Cape Verde and São Tomé, Guinea, Angola and Mozambique, to which was added the fortress of São João Baptista de Ajudá, a detached portion of São Tomé e Príncipe. In Asia, the State of India, comprising the territories of Goa, Daman and Diu, and the administration of Macau. In Oceania, the eastern half of the island of Timor. It was the fifth largest empire among the European powers, yet Portugal was far from controlling the near two millions square miles that its empire encompassed.

A heirloom from the imperial model – informal and based on trade – of the old regime, despite all occupation efforts since 1875, Portuguese sovereignty remained mostly centred in coastal areas and in a few enclaves in Angola and Mozambique. The transition to a modern imperial model, based on economic production and in territorial control and administration, was a slow and drawn-out process extending throughout some thirty years.

The process of creating a colonial State required, in a first stage, the territorialisation of the empire, which was carried out by means of numerous military campaigns – the so-called "pacification campaigns". They had begun in the previous decade, but lacked the military nature they thenceforth acquired. They lasted from the nineties until the First World War, and unfolded in two distinct periods: the first, during the Constitutional Monarchy, between 1890 and 1909, focused in the occupation of key zones in Angola and Mozambique; the second, contemporary with the Republic, between 1910 and the post-First World War, centred in the peripheral provinces of Guinea and Timor.[44]

Although numerous, those military campaigns were generally small-scaled. Their tactical goals were the occupation of territories and the subjugation of relatively small populations. They deployed a short number of men, for the most part indigenous troops, but enjoyed military superiority, given the armament at their disposal comparing to the African resistance fighters. Sometimes, and they were few, when strategic goals and political outcomes were at stake, that is, when Portuguese sovereignty or the consolidation of the Army were at stake, the military campaigns took on a larger scale, with the deployment not only of colonial troops, but also of metropolitan forces and the use of significantly greater fire power. This was the case, for instance, in the south of Mozambique, with the Gaza campaign against the *vátua* (ethnolinguistic group that inhabited Southeastern Africa) State and the capture of its sovereign, the notorious Gungunhana, which attained proportions of heroic feat and, in the eyes of the imperial political imaginary, exacted some kind of symbolic revenge for the Ultimatum.

Mozambique

Mozambique represented, in any case, top priority in the "pacification" campaigns. The reason behind it was simple and ensued from the ill-fated Pink Map: it was imperative to occupy the territory south of the Save River, coveted by Cecil Rhodes, whose pressure was relentless. After sovereignty over south Mozambique was consolidated, new military campaigns would be launched towards the north. They began as soon as 1895 – conceived by António Enes, who was the new Royal Commissioner and endowed with full powers – and were carried out by already renowned army men, therefrom turn into heroes, such as Mouzinho de Albuquerque, Ayres de Ornelas and Paiva Couceiro.

The campaigns unfolded in three different phases. The first, in the south, was precisely the Gaza campaign against the *vátua* State. Following the battle of Marracuene, in February of 1895, in which Portuguese troops succeeded in suppressing the last pockets of Tsonga resistance, the region of Lourenço Marques was therefore "pacified", and strategic conditions created in order to launch the Gaza campaign. Politically, the campaign also intended to prevent the *vátuas* from allying with Cecil Rhodes and serving the interests of the British South African Company. Moreover, the moment was favourable, as the *vátua* State was going through a series of crises and militarily weakened by a combination of circumstances such as internal uprisings, an outbreak of epidemics and the mass migration of young men to work in the South African gold mines. The campaign took on big proportions, deploying an expeditionary corps of 2,200 soldiers from the metropolitan army and considerable fire power.[45] The first strategy did not involve direct confrontation and some voices even argued for a peaceful solution and the establishment of a protectorate over Gungunhana's empire. That was the ultimate objective of neither António Enes nor the military, let alone that of the government in Lisbon. Strategically, circumstances advised direct confrontation and a military offensive, and the operation plan was drawn up accordingly. After Paiva Couceiro and Freire de Andrade's victory at the Battle of Magul, on the 7th of September of 1895, the Portuguese troops headed north and, in November, advanced into the interior of the Gaza kingdom. On the 28th of December, Mouzinho de Albuquerque, in a bold raid over Gungunhana's kraal, captured the king of Gaza. Rumours had it that António Enes had promised he would take Gungunhana, defeated and made prisoner, to the royal presence. Truth or not, the fact is, once captured, the *vátua* chief was transported to the metropole and, in March of 1896, displayed in Lisbon as a trophy. Finally, he was sent into exile in the Terceira Island. The victory over the *vátuas* and Gungunhana's capture marked the end of African resistance to Portuguese sovereignty in the south of Mozambique. The military achievement was not devoid of political meaning: it was invoked as proof of Portugal's capacity to uphold its imperial project.

The south being "pacified", the campaigns progressed to the north and to the centre. In the centre, the objective was Zambezia. Unlike what had occurred in Gaza, where campaigns were brief and victories conclusive, the conquest of Zambezia was a slower and more difficult process. There were many small localised campaigns, through which sovereignty over Majanga da Costa, in the littoral, and over Angónia, which borders on Nyassa and Macanga, north of Tete, was gradually established. In Zambezia, the great military campaign was that of Barué, in 1902. Strategically, Barué was an asset for Portugal. Not alone for being in itself a territorial unit, but also because its geographic location on the Rhodesian border endowed it with strategic value. For that reason, the slightest disruption could mean trouble with England and endanger Portugal's credibility as a colonial power. That is, question Portugal's effective capacity to occupy the territory and exercise sovereignty. Now, following a failed attempt at indirect control of the territory through the mediation of the Mozambique Company, the State was forced to rule directly. A sizable military campaign was organised, deploying a force of around 1000 regular army men – half of which came from the metropolitan army – and a corps formed by 15,000 *sepoys*. Equipped with modern armament, it was commanded by João de Azevedo Coutinho.[46] The contingent involved and the weapons used, especially the machine gun and the artillery, conferred to the Portuguese forces a significant military superiority that, within three months, enables them to seize control of the territory and subdue the populations. Azevedo Coutinho thus sealed the "pacification" of central Mozambique. Not conclusively, however. During the First World War, a new uprising challenged Portuguese sovereignty, calling for a new fairly-sized military campaign. Resistance was finally defeated in the following year, and Zambezia remained "pacified" until the war of decolonisation.[47]

Similarly, the "pacification" of the north was neither a swift nor an easy process. Since 1893, the whole region between the Rovuma and Lúrio Rivers had been concessioned to the Nyassa Chartered Company, which ruled on behalf of the sovereign state yet possessed no great means, ultimately failing to rule the territory in any effective way. On the other hand, Mouzinho's military incursions, in 1896, against the Namarrais in the Macua zone, were far from successful. This meant that, until the first decade of the 20th century, not only commercial networks and slave trading persisted between the peoples of the interior and the Muslim Sheikhs of the littoral, there were vast zones and populations that eluded Portuguese sovereignty. It was only in 1908 that the Nyassa Chartered Company was restructured and begun to exercise control over the colony. In 1912, it defeated Mataca, the chief of the Ajaua, and begun to establish administrative and fiscal control of the Nyassa. Since 1905 and later on, during the Republic, several military campaigns were launched in the littoral south of the Lúrio River, namely in Angoche, gradually expanding towards the interior. Still, all this "pacification" moves in the north were stalled by the beginning of the Great War in the African theatre, which affected directly the whole north of Mozambique. There, on opposite sides of the Rovuma River, the Portuguese army fought the German troops, and only the Versailles Peace Conference, in 1919, recognised conclusively Portuguese sovereignty over the territory south of the Rovuma.[48]

Angola

Except for the absence of chartered companies, the "pacification" of Angola followed a similar pattern to that of Mozambique. By and large, it occurred gradually over time, involving more than 180 small campaigns – modest in their objectives and means. In some cases – not many – it required larger military campaigns, whenever the need arose to occupy strategically important territories or to defeat more recalcitrant insurgencies against Portuguese sovereignty. Such was the case with the campaigns led by Alves Roçadas between 1904 and 1907, in the south against the Cuamato, and the ones led by João de Almeida, between 1913 and 1914, in the Dembos. Other instances are the campaigns against the revolt of the *ovimbundu* people in the Bailundo war, in 1902, in the central plateau, and later on, between 1913 and 1914, and against the 1913 uprising in Congo. In these cases, in addition to the indigenous troops, metropolitan troops were deployed and more powerful and modern weaponry was used.

In the north of Angola, after the Berlin Conference had allotted Portugal the Cabinda enclave and the left bank of the Zaire River, the region underwent a reorganisation, with the establishment of military, administrative and fiscal stations, and the occupation expanded towards the southern littoral, including Ambrizete (1886) and afterwards, to the interior, encompassing Maquela do Zombo (1896). The objective was, from there, to go further into the hinterlands and extend sovereignty onto the border with the Congo Free State. In truth, however, except for a "residence" in São Salvador, former capital of Congo, Portuguese penetration into the interior was little more than formal. A little further to the south, in the region of the Lundas, after the expeditions led by Henrique de Carvalho between 1884 and 1887 and Cândido Sarmento in 1890, the question of the Angola/Congo Free State frontier was settled by diplomatic means in 1891. Portugal's sovereignty claims over the territories between the Congo and Cassai Rivers, and Leopold II's over inland territories, was recognised. The border having been defined, the occupation progressed, since 1892, in a slow and gradual fashion, with the establishment, between 1886 and 1903, of military and fiscal stations, under the government of Cândido Sarmento, with Malanje as the epicentre. Only later, between 1907 and 1909, when Paiva Couceiro served as Governor of Angola, was a new military offensive into the interior launched. Although fragile, because it was based on a scant network of stations relying on flimsy communications, the occupation of the northeast was completed.

In the south, the situation was altogether different, and since 1886, following Artur de Paiva's campaign, military positions had been established in the southwest line of penetration: Princess Amelia and Maria Pia Fortresses. Efforts, however, centred in the border with South West Africa, where Portuguese sovereignty collided with German rule. The epicentre of Portuguese sovereignty was the Huíla Plateau, where white colonies had been settled since the eighties. The first great military operation beyond the Cunene River, with the purpose of occupying the Cuamato and Cuanhama areas, was only launched in 1904. Under the command of Alves Roçadas, a contingent of around 2000 men was deployed, approximately half of which being metropolitan troops. But the Portuguese offensive was fended off as soon as it entered Cuamato, in September of 1904, and having suffered heavy losses,[49] it was forced to withdraw. This military fiasco, which became known as the "Vau do Pembe disaster", led to the deployment of a new expedition in 1906, engaging more soldiers and well equipped with modern weapons, which crossed the Cunene onto its left bank and, in 1907, finally made it into Cuamato. After fierce combats, the machine gun and the artillery confirmed the military superiority of the Portuguese, which seized the *embalas* (seats of power) of Cuamato Pequeno and Cuamato Grande. The expedition retreated, but five Portuguese fortresses were built in the region, one on the left bank of the Cunene and four in Cuamato. In subsequent years, João de Almeida endeavoured to consolidate Portuguese sovereignty in the region, but the *cuanhamas* invariably resisted. On the eve of the Great War, as was the case with the border north of Mozambique, Angola's southern border was precarious, and so the region was the stage of a number of military operations between the Portuguese and the German since 1914 and until the end of the war, when the Versailles Conference recognized Angola's southern border and Portugal's colonial integrity.

The third front in the occupation of Angola was the central plateau. In 1890, a campaign led by Artur de Paiva had defeated the *soba* (village chief) of Bié and installed the Belmonte Fort, thus establishing Portuguese sovereignty in the region. The following year, the same took place in Bailundo. But the thin military presence of the Portuguese and a strong resistance by the Africans rendered frail the Portuguese rule, which was ceaselessly defied by insurrections and revolts. The revolt of the Ovimbundu in 1902 was the most consequential. It all began with a simple refusal to pay off a debt that the Captain-major of Bailundo had claimed. Against the backdrop of latent tension between rulers and ruled, however, the imposition of the Portuguese authorities together with the *soba*'s refusal to make an appearance at the fort led to the imprisonment of the latter in May of 1902. It was all it took to trigger a confrontation between the garrison and the local population, which threatened to invade the small fort. The Bailundo people called for allies and reinforcements arrived from Bié and Huambo. The tension escalated and the armed conflict became inevitable. The Portuguese authorities were forced to ask for reinforcements and a large armed force, equipped with machine guns and artillery, was deployed. Not only did the Ovimbundu besiege the fort, they also attempted to cut off communications between Bailundo and the coast, in order to halt the Portuguese reinforcements. They succeeded in taking the fort, but did not manage to cut off communications. And when the Portuguese convoy arrived, their military superiority settled the matter. The siege was dismantled and the fort was freed on the 10th of June. The Portuguese military forces then began their offensive and seized the Ovimbundu *embalas*. The skirmishes continued, with successive victories by the colonial troops. On one such occasion, on the 4th of August, the military chief at the head of the revolt, Mutu-a-Quebera, is killed in combat. Lacking its leader, the revolt waned and resistance lowered its arms. In October, the "pacification" of the Bié Plateau was concluded.

A sizable revolt against Portuguese rule still broke out in Congo in 1913. Similar to the Bailundo revolt, it required a larger military campaign, which lasted for a number of months until the capture of the rebellious leader, Álvaro Buta, in August of 1914. Other minor uprisings, scattered and sporadic, went on throughout the years of the war, which in the meanwhile changed the pattern of military occupation.[50]

The minor theatres

Besides Angola and Mozambique, two other – in this case, peripheral – overseas possessions were the subject of "pacification" campaigns: Guinea and Timor.

In Guinea's case, the delimitation of borders had been settled with France in the Portuguese–French Convention of 1896, but effective occupation of the interior of the territory was practically inexistent, and Portuguese sovereignty was limited to three villages in the littoral: the Bolama Island, which was the regional capital, Cacheu in the mouth of the homonymous river, and Bissau in the mouth of the Geba River. During the "time of the Centurions" (1890–1909), the "pacification" of Angola and Mozambique was top priority, hence military means – human and material – available in Guinea were scarce. From the eighties, the garrisons were gradually reinforced with troops deployed from the metropole and other colonies, and with the artillery and naval equipment that the geographic nature of the territory demanded. By virtue of Guinea's ethnic diversity, the campaigns were based on those scarce military resources and on a policy of unstable alliances with one ethnic group to subdue another, as was the case with a number of revolts against Portuguese colonial rule. Actually, it was not before the "times of the clearances" (1910–1920), under the command of Major João Teixeira Pinto, between 1912 and 1915, that an investment was made in a genuine policy of territorial occupation for the hinterlands. Throughout that period, taking advantage of their military supremacy and the alliance with the irregular forces of the Cuor chief, Abdul Injai, the Portuguese colonial forces launched a series of campaigns that succeeded in defeating and subduing the various Guinean ethnic groups, seizing control of their territories: the *Balantas* and *Mandingas*, in 1913, the *Manjacos* and once again the *Balantas*, in 1914 and, finally, the *Papéis* and the *Grumetes*, in 1915. This process was accompanied by the installation of military and administrative stations, as well as a tax collection system, and, later on, the construction of a network of roads and communications. As was the case with other territories, however, the "pacification" of Guinea was not accomplished without resistance to the gradual foundation of a new order.

In Timor, the pattern of territorial occupation did not deviate significantly. Here also, by the end of the 19th century, Portuguese sovereignty was circumscribed to a few points in the littoral: the capital Dili and a handful of military posts, seven in the northern coast and three in the southern coast. The whole of the interior was the realm of its tribal chiefs. Effective occupation was undertaken by means of a series of campaigns launched during the government of Celestino da Silva, and here also the contingent was small and the creation of alliances with tribal chiefs a common resource.

With either the purpose of establishing sovereignty or that of squashing revolts against sovereignty, "pacification" campaigns were a crucial instrument in the occupation of the territories and the subjugation of its populations, that is, in the territorialisation of the empire and the construction of the colonial State. As in every colonisation process, the use of force was a reality in the Portuguese case. However, it was used with restraint and there were no extermination attempts. Granted, perhaps the military capacity – or lack thereof – of the Portuguese did not advise it in any event, and that is why the alliances with local chiefs had a decisive part. Not just in the management of military operations against other ethnic groups, but also as insurance regarding administrative control, workforce recruitment and effective tax collection after the "pacification". "Pacification" that only the Great War and the Versailles Conference finally consolidated.[51]

CHAPTER

3

The Republic and the Great War

The international standing of the Republic before the war

Towards the end of the Monarchy, Portugal's international standing was fragile and Portuguese foreign policy struggled with two crucial questions on two different fronts: the colonies and the Iberian Peninsula. The common denominator was the Anglo-Portuguese Alliance.

On the colonial front, it concerned the growing Anglo-German rivalry and the use of colonies as a mechanism for counterbalancing and levelling the international scene, in the late 19th century.

Ever since Germany emerged in the seventies as a great power in the European scene, and, from the nineties, as a naval power with colonial pretensions, relations between England and Germany had a decisive impact on international relations. Anglo-German rivalry and its management were crucial to the stabilisation of the international scene, and in the relationship between the two powers the colonial issue played a significant part. That is, to ensure the stability of European balance, England was often forced to yield to Germany in colonial matters. Colonies in general and Portuguese colonies in particular served as a bargaining chip in this scenario. International historiography has debated at length the role of the colonial issue in Anglo-German relations: were the colonies a strategic goal *per se* or merely a tactic of appeasement and bilateral rapprochement?[1]

Strategic goal or tactic of appeasement, in any case politically it meant the same for Portugal: a threat to the territorial integrity of the African empire. This threat had loomed as early as in 1898, when the Monarchy was still in force, and now, on the eve of the war, in 1912–1913, with the Republic in full swing, it was once again impending: the Anglo-German agreements over the partition of Portuguese colonies. A threat that would weigh decisively on Portugal's decision to enter the Great War.

The 1898 agreement should be understood in the light of two convergent dynamics: on one hand, internationally, the colonial panorama of the second half of the nineties, namely the imperial projects of the rival powers, England and Germany; on the other, domestically, the difficult financial situation of the Portuguese State, which was declared bankrupt in 1892 and deteriorated all through the nineties.

In an effort to reverse the situation and consolidate public finances, Portugal pondered asking for a loan from London using the revenue from African colonies as a security. Upon receiving news of negotiations between England and Portugal over a possible loan, Germany let the British government know of its interest in the operation. The conjuncture appeared to be favourable to an Anglo-German rapprochement and the proposed loan to Portugal was immediately seized as a pretext to open bilateral negotiations between Germany and England. These negotiations led to the signing of an agreement that stipulated two basic clauses: that the loan should be secured by the customs revenues of Portuguese colonies and at once specified the partition of revenues between England and Germany. In accordance to their respective strategic interests, England would have Mozambique and Germany would receive Angola. The second clause established that, in the event of a default on the loan, each zone would automatically be integrated in each country's respective "sphere of influence".

However, for the agreement to hold, two conditions were required: in the first place, that Portugal failed to service its debt; and secondly, that both powers were equally willing to carry it through. Neither one took place. And, despite having been discussed in secrecy, the terms of the agreement seeped out to the Portuguese Embassy in London. The Ambassador informs the government and Lisbon declines the loan. On the other hand, whereas Germany was adamant in wanting the agreement closed and published, the same could not be said of England, which would see exposed, not only its diplomatic duplicity, but also the blatant contradiction between the interests behind the Anglo-German agreement and the obligations of the Anglo-Portuguese alliance. An ambiguous solution would suit England best: signing the agreement that would neutralise Germany, yet not closing the deal that would estrange Portugal.[2]

The onset of the Anglo-Boer War and the evolution of the international conjuncture will eventually benefit the Portuguese position.[3] Aware of the strategic importance of the Lourenço Marques's port and railways, England will ask Lisbon for diplomatic and logistic support in using these resources for their war efforts. Having knowledge of the

Anglo-German agreement, and faced with London's diplomatic request, Portugal sets in motion the so-called "diplomacy of opportunities". Taking advantage of the situation, Portugal plays its card, showing willingness to grant its support to England, but on one clear condition: the reaffirmation of the Anglo-Portuguese Alliance in the terms of the 1642 and 1661 treaties. In fact, this would be the origin of the secret declaration of 1899, ratified and made public by the signing of the Treaty of Windsor in 1904. The Anglo-Portuguese Alliance reaffirmed, a new period began marked by a renewed friendship between Portugal and England, which lasted throughout the last phase of Dom Carlos's reign.

However, in the last days of the Monarchy, the new challenges and threats faced by Portugal and Portuguese foreign policy will arise within Europe and the Peninsula.

Ever since the enthronement of Alfonso XIII, in Spain, the neighbouring power had relinquished its traditional isolationism and launched a diplomatic rapprochement with the powers of the *Entente Cordial*, especially with Great Britain. After the Treaty of Algeciras, the need for naval cooperation from Spain had become clear for England, concerned with ensuring status quo in the Western Mediterranean. The cooperation reinforces the diplomatic rapprochement between Edward VII's England and Alfonso XIII's Spain, embodied in 1907 by the Pact of Cartagena.

Cartagena became thenceforth a spectre looming over Portugal – the shadow of the "Spanish peril". Spain's rapprochement with the same system of extra-peninsular alliances, and especially, the rapprochement between the "old ally" and the "traditional threat" disrupted the fragile Portuguese geopolitical balance and magnified drastically the vulnerabilities of national security: not only because they rendered the Portuguese territory strategically less significant in the peninsular framework and politically devalued Portugal's position vis-à-vis Spain, but above all because they undermined the value of the alliance as an instrument of defence and guarantor of national independence.[4] It was feared that national independence might ultimately be at stake.

If England had already yielded to Germany in colonial matters, why would it not yield now to Spain in the context of the Peninsula?

Bound by the Anglo-Portuguese Alliance, this was the core fragility of Portugal's international standing and the great challenge faced by Portuguese foreign policy by the end of the constitutional Monarchy. And although kindred regimes and the dynastic kinship between Dom Carlos and Edward VII had mitigated the risks of Cartagena in 1907, the regicide in 1908 represents an abrupt disruption in the course of events.

The Republic will abide by the same principles that had guided foreign policy during the Monarchy, inherit the same position of international weakness and face, now in the absence of regime and dynastic kinship, the same hardships and the same challenges.

The international conjuncture was in no way favourable to the implantation of a Republic which, moreover, would take on a progressively more radical and Jacobin character. The Europe of the time was predominantly monarchic and the only two other republican regimes did not represent significant sources of international support, let alone alternatives for Portuguese foreign policy – Switzerland was a small power devoid of weight in the international scene and France was diplomatically tied to England by means of the *Entente Cordiale*.

When, on the 5th of October of 1910, the Monarchy is overturned, the new Republic is greeted internationally with a degree of aloofness that soon became outright hostility.

Earning official recognition will represent the first great international challenge faced by the Republic. It would prove to be a difficult and complex process taking place in three distinct waves and dragging for ten long months, with grave consequences for Portugal's international situation and domestic policy.

The first wave of official recognition of the Republic is almost immediate and comes from the South American republics, topped by Brazil and Argentina.

The second wave, between June and August of 1911, was led by the Republic of the United States of America and the French Republic. The deferment of recognition had different causes in one case and the other. In a breach of the Monroe doctrine, which stipulated the *de facto* recognition of regimes, it repeatedly delayed and postponed the recognition of the Portuguese Republic. Washington was concerned with assuring the constitutional lawfulness of the regime in order to preserve a degree of diplomatic caution vis-à-vis the recognition of the multiple Latin-American dictatorships. France, bound to England by the *Entente Cordiale*, followed the latter's positions and made its recognition conditional on all the political-constitutional assurances that the London government had demanded of the republican regime.

Recognition from the two powers came when all the required formal guarantees had been met by the Portuguese Republic, although recognition from England was still overdue.

The last wave of recognitions was led by the British Monarchy, which was accompanied by the all the other great European monarchies. London made its recognition dependent on a number of conditions that kept postponing it: first, the holding of elections and the establishment of a proper functioning Constituent Assembly. After the Assembly had been elected, the election of the President of the Republic. And when the head of State had been elected and was exercising his duties fully, England kept delaying its recognition. The reason behind it was not in fact juridical-constitutional, but of a political nature, and it had to do with the litigation between the Portuguese Republic and the Anglican Church in Portugal, following the Law on Separation of Churches and State. This situation was only unblocked on the 11th of September of 1911, when England and the other European monarchies jointly expressed their recognition.

Official recognition was undoubtedly a first step, but it did not mean full international acceptance of the regime.

The implantation of the Republic did not alter the course of foreign policy or the strategic options of the Portuguese State, namely its commitment to the Anglo-Portuguese Alliance and the African colonial project. But neither did it change the threats and challenges that weighed on the country – the Republic will go on struggling with the same problems faced by the Monarchy, but in a more fragile position, as internationally the situation worsened.

Despite its international isolation, the Portuguese Republic still manages to take part in the dynamics in motion in the international scene, as much in Europe and in the Peninsula, as overseas in the colonies.

In Europe, in the Mediterranean and the Atlantic, Portuguese interests and those of the European powers still intersect.[5]

In the Atlantic, the Azores remains the focus of ever-growing attention and interest, and the change of regime has no impact on England's position in this regard. The strategic value of the archipelago, which was clear even in the late 19th century, increases in the same measure as the naval rivalry between England and Germany in the Atlantic. And despite the fact that, in 1912, a report from the Admiralty questions the value of the Portuguese insular territory, interest for the Atlantic Islands never wavers and the request for assurances on the exclusiveness of facilities is renewed by London with every new Foreign Affairs minister in Portugal.

In the Mediterranean, the Portuguese presence is marked by its accession to the Franco-German agreement of 1911 which put an end to the Agadir crisis. Portugal, bound to support the powers of the *Entente Cordiale*, could not however overtly oppose Spain, which had yet to solve its differences with France. The difficult relations between the Peninsular states requires of Portugal a subtle diplomatic performance – successfully carried out – that involves postponing the accession until after the Franco-Spanish agreement is signed, in May of 1912.

The spectre of the "Spanish peril", however, once again hangs over the Peninsula. And the Portuguese diplomatic ability in the Mediterranean process notwithstanding, the Spanish rapprochement with the powers of the Entente, England in particular, revives the ghost of Cartagena and brings to the fore the central issue of Portuguese foreign policy – the geopolitical balance between Spain's continental pressure and the maritime compensation of the Anglo-Portuguese Alliance.

The annexationist lobby in Spain was well known and, despite not representing the official position of Spanish foreign policy, the idea was not at all dismissed by Alfonso XIII himself. Moreover, the atmosphere of political instability in Portugal leveraged and facilitated this hypothetical scenario. In such conditions, would the Anglo-Portuguese Alliance prove to be an effective enough diplomatic and military instrument to ensure Portuguese security?

This was a sensitive question and the Foreign Office chose not to interfere directly in the peninsular issue, encouraging instead a bilateral understanding between the parties. In the tensest moments, however, it was forced to take a stand: it did not go so far as to condone the military intervention in Portugal and the country's annexation, but still it dealt with Spain with exceptional tolerance during the monarchist incursions. On two occasions, in 1911 and 1912, starting from Spain with the knowledge and approval of the Spanish government and led by Captain Paiva Couceiro, the monarchist troops marched into the Portuguese territory with the unconcealed purpose of restoring the Monarchy. The incursions met with military failure, but had the immediate effect of building up the spectre of the "Spanish peril".[6]

In light of this, the Lisbon government repeatedly attempted, between 1912 and 1913, to obtain formal assurances from the Foreign Office regarding the terms of the alliance. To no avail. Oral reassurance is all it managed to obtain.

The reason was simple enough and had to do with the other feature of Portugal foreign policy: the colonial question.

Similarly to what had happened in 1898, the years of 1912–1913 will witness the signing of a second Anglo-German agreement on the partition of the Portuguese colonies. Like the first, this second agreement can also be explained by the convergence of two distinct dynamics: on the one hand, the strained international conjuncture, the arms race and the Anglo-German rivalry before the war; on the other, the difficult post-revolutionary situation in Portugal, marked by international isolation, and the Republic's unstable domestic policy.

Following the resounding failure of the Haldane mission, the last attempt at reaching a diplomatic agreement between England and

Germany, in 1912, the only remaining point of contact for dialogue between England and German was the colonial question. Awarding Germany compensation overseas appeared to be England's last resort to maintain the already tottering balance in Europe and to avoid conflict.

The internal and external vulnerabilities of the Portuguese Republic worked in favour of this approach. In 1912, the two powers signed a treaty renewing the terms of the never enforced yet also never denounced 1898 agreement: it renegotiated the partition of spheres of influence and, above all, introduced a clause that broadened the legal basis for an intervention, by the signatory powers, in the Portuguese territories. Like the former, this second agreement, under the autonomous but simultaneous pressures of Portuguese and French diplomacies – the latter excluded from the partition – was also unsuccessful. The onset of the Great War decided its definite impracticability. One thing was certain, however: that instrument of international stabilisation, which England's foreign policy never shied from wielding whenever it appeared to serve its interests and objectives, represented a real threat to the territorial integrity of the colonies and to the Portuguese imperial project.

When the First World War broke out, the international standing of the Portuguese Republic was extremely fragile, threatened in the continent by Spain, in the colonies by Germany and made twice vulnerable on both fronts, because Portugal's sole instrument of foreign policy – the Anglo-Portuguese alliance – was carrying out a policy of complaisance in the two scenarios – with Spain regarding the peninsular question and with Germany regarding the colonial question.

These two factors, along with the domestic political situation, ultimately decided Portugal's entry in the Great War.

Entering the War: National objectives and political strategies

Why did Portugal enter the Great War? The question is straightforward enough, but the answer is rather a complex one. Portugal's position regarding the war did not remain unchanged from 1914 to 1918. Much to the contrary.

Diplomatically, it evolved from unstated Anglophile neutrality in 1914, under more moderate coalition governments, to an attempt at equidistant neutrality, in early 1915, under the dictatorship of Pimenta de Castro, and, further ahead, from the revolution of the 14th of May and under more radical governments, to a path of belligerency, until entering the war in March of 1916.[7] Even after the entry, the war policy does not remain constant, but reaches another turning-point between 1917 and 1918, during the government of Sidónio Pais. Obviously war in Africa, which gathered consensus in Portuguese society, was one thing; the war in Europe, which opened up the greatest divides, within and outside the political system, was quite another. In the framework of political instability of the Republic, a number of governments fell because of the war and others came into power to change the war policy. All of them wielding different objectives and different strategies, thus depriving Portugal of a single national strategy.

Also from the military point of view, one reality was the African theatre, where skirmishes with Germany took place in the south of Angola and the north of Mozambique, from October of 1914; another completely different reality was the European theatre on the Western front.

There is no simple answer for this question, but historiography has so far come up basically with three different explanatory theories – the colonial thesis and the European-peninsular thesis, which correspond to the two international factors that have traditionally been deemed more relevant; and a third explanatory theory connecting to those external factors the internal political question.

The colonial thesis, dating back to the first line of the legitimisation strategy of the official political discourse of the time, evolves in historiography between the two wars and extends, in specific trends, to the present day. In a nutshell, this thesis states that Portugal entered the war in order to save the colonies.[8]

This is unarguable. As we have seen, the Portuguese colonies in Africa were the object of the economic and strategic covetousness of the great powers: France, England and Germany. Furthermore, they were able to be used – and indeed they were, more than once – as a compensation mechanism and bargaining chip in the European balance of powers. On two occasions prior to the war, in 1898 and in 1912–1913, England and Germany secretly dickered between themselves the partition of the Portuguese overseas possessions. During the war, a number of powers showed themselves interested in their fate: France, Belgium and even Italy. But obviously, above all others, Germany and England. England, who carried out direct and military attacks against the Portuguese colonies and orchestrated uprisings among the indigenous populations against the Portuguese rule; and Germany, who made strategic use of ports and territories for the logistic support, disembarkation and passage of troops in their war efforts in Africa. Moreover, English resistance to Portugal's entry in the war also had to do with the British War Cabinet's wish to avoid not only possible territorial claims by Portugal in the post-war colonial framework, but also being bound by any obligation regarding those same colonies, so they could be played as trump cards on the negotiating table should the outcome of the conflict made it necessary.

There is no doubt whatsoever that the Portuguese colonies and Portugal's future sovereignty over its overseas territories were under threat. The colonial question had a decisive weight and represented a crucial mobilising factor, indeed the only one that gather consensus in Portuguese society. To enter the war alongside the Allies and with England's blessing was an additional assurance. Assurance, to begin with, against Germany's territorial claims in southern Angola and northern Mozambique. Assurance also against British urge to use the Portuguese colonies as a compensation mechanism in peace negotiations.

The colonial thesis is thus a true, if incomplete, thesis. It explains partially, but does not fully account for Portugal's entry in the war, and it certainly does not explain the type of military intervention, the hierarchy options and the selection of operational theatres. That is, the defence of the colonies did not justify intervening in the European theatre, let alone active belligerency. Colonial sovereignty and integrity might have been safeguarded by means of a different strategy and by opting for a different operational theatre, that is, channelling the war effort and directing military intervention exclusively to the African theatre, which being peripheral and secondary to the *enjeux* of the global war, did not require Portugal to resort to active belligerency. England, Germany and the more moderate republicans in Portugal knew it perfectly well. England never saw the services rendered by Portugal, not even military cooperation in Africa, and the preservation of neutrality as irreconcilable. Indeed it found Portuguese neutrality advantageous insofar as it facilitated the provision of those services. Germany, in its turn, despite all the violations of neutrality and the military conflicts in Africa which were, after all, war-waging against Portugal, did not find in it a good enough reason for a declaration of war until 1916. The moderates were aware of this from the start and made it their strategy: to intervene militarily in Africa while maintaining discreet and unofficial cooperation with England, that is, tacit Anglophile neutrality. The colonies were not, therefore, sufficient grounds for Portugal's entry in the European war.

The second explanatory theory for Portugal's entry in the war is essentially based on the European-peninsular thesis.

More recent in historiographical discourse,[9] this theory evolves mainly since the 70s and 80s in Portugal and Spain. It retrieves a second, less emphasised, legitimisation strategy of the official political discourse of the time – the country's international prestige and the attainment of a place in the concert of nations – while adding to it a new element wherein, in fact, the crux of the explanation resides: the peninsular question. In other words, the issue of the "Spanish peril". In short, Portugal entered the war to accomplish two objectives: the first one, vague and remote, was to reconquer its lost place in the concert of nations at the European level; the second, concrete and immediate, was to dispel the "Spanish peril" in the peninsular context.

This point is also beyond contention. Before the war, Portugal's international situation was a difficult one and it did not become any easier in the course of the conflict. Maintaining some sort of balance between the continental pressure exerted by neighbouring Spain and the overseas compensation of the Anglo-Portuguese alliance was a hard task. In the first decade of the 20th century, the conjuncture deteriorated further: England, Portugal's old ally had become diplomatically closer to the historical threat that was Spain, which weakened the alliance and dangerously reduced the guarantee over the security of Portuguese territorial integrity and perhaps even Portuguese sovereignty. If the spectre of Cartagena hovered menacingly since 1907, that is, since the Monarchy, the implantation of the Republic in Portugal made relations between the peninsular states considerably worse. Before the war, with the knowledge and approval of the Spanish Monarchy, movements set on restoring the monarchist incursions of 1911 and 1912. During the war, there was a generalised will for annexation in Spain and the annexationist lobby was powerful. Bilateral relations between Portugal and Spain were invariably difficult and the balance unstable. The "Spanish peril" was real and the concern it elicited was all the more justified as England had advertised to Portugal on more than one occasion their restrictive interpretation of their obligations under the alliance: they would guarantee the defence of the Portuguese shoreline and colonies, but not the land border.

Although it was never fulfilled, the Spanish peril was palpable and not only did it condition Portugal's political and diplomatic conduct, it was also a topic for mobilisation and propaganda in favour of a Portuguese intervention.

In the face of Spanish neutrality, to enter the war alongside the Allies, by the hand of England, meant twice the assurance.

Assurance, in the first place, within the peninsula: entering the war alongside England, while Spain remained neutral, was a way at once to weaken the Spanish-British rapprochement and to reinforce the Anglo-Portuguese Alliance. On the other hand, it meant a strategic diversification of the Portuguese territory and an affirmation of Portugal's prominence in the Iberian Peninsula. Lastly, it meant ensuring the position of privileged partner and preponderance in the peninsula. In other words, fending off, once and for all, the "Spanish peril".

Secondly, assurance within the European framework: by entering the war side-by-side with the Allies, Portugal hoped to earn, after the conflict was over, a place in the concert of nations and the international recognition which, ever since the implantation of the Republic, it had been granted *de jure*, but was still lacking *de facto*.

Like the colonial thesis, the European-peninsular thesis is also a true thesis. Not only that, it is undoubtedly more elaborate than the former. It is still an incomplete thesis, however. If, on the one hand, it accounts for the belligerency, it fails to fully explain the choice of operational theatre. There is no doubt that eradicating the "Spanish peril" required that Portugal took on a greater prominence in the peninsula and conquered a premium position in extra-peninsular alliances, that is, in its relationship with England. It is also unarguable that, in order to achieve these objectives, differentiation in the international status of the country was essential – neutral Spain required a belligerent Portugal. The need for belligerency was thus established, but the option for a military intervention in the European theatre rather than in Africa – and this was precisely at the heart of the controversy – remained unexplained. The exercise of belligerency could have been limited to the diplomatic victory, military cooperation in peripheral theatres and the provision of services to the Allies – services that Portugal had always provided, but now with the advantages of the new diplomatic status. Such was indeed the intention of the British Cabinet when it decided to invoke the Anglo-Portuguese Alliance to requisition German ships stationed in Portuguese ports. A belligerent stance did not entail intervening militarily in the European theatre. What is more, that intervention added very little to the already achieved national goal. Notwithstanding, Portugal marched towards the trenches of Flanders a few months later. Now, the question is that peninsular affairs do not suffice to explain, *per se*, the military intervention in the European theatre. And why? Essentially, because the political purpose of Portuguese interventionism was not exhausted by external reasons.

Despite their differences, the colonial and the peninsular theses rely on a common assumption: their explanatory theories are based on factors and objectives pertaining to external policy – the colonial question in the first instance, the peninsular question in the second, and the Anglo-Portuguese Alliance in both cases.

Not only do they overlook internal factors, they also disregard their weight on the evolution of foreign and defence policy itself. That is, they dismiss the fact that war goals and political strategies lack national consensus and, as such, vary according to the evolution of domestic policy. They dismiss ultimately that the explanation for radical strategy, the triumph of interventionism *à outrance* and the option for the European war theatre very likely have domestic motivations.

The above means that both theses are necessary but insufficient to answer satisfactorily the question of why Portugal entered the Great War.

The third explanatory hypothesis is more complex and multilevel, and adds to external constraints issuing from domestic policy.[10]

Ever since the implantation of the Republic, Portugal had been struggling with a structural and never sorted out political problem: political consolidation and national legitimation of the regime.[11] Since its onset, and resorting to juridical and political mechanisms, the republican regime thwarted access to political participation by important sectors of Portuguese society, left and right of the party spectrum. Left out of the legal political framework, these sectors – especially royalists and Catholics to the right and labour and union movements to the left – often embark in illegal forms of political representation, from public disturbances to revolutionary attempts, thus increasing the levels of political violence. When the regime fails to integrate the whole of the Portuguese, the Republic is not only deprived of national legitimacy, but is also constantly under threat. One other, no less serious, issue troubled the Republic from within the political system: internal dissension and irreconcilable rivalries among republican parties, both radical and moderate. The problem of governability and of the regime's political stability was hence added to that of consolidation and legitimacy.

The Republic was therefore torn in its very political structure by profound divisions which the war aggravated and overdetermined: outside the regime, royalists were divided into Anglophiles and Germanophiles, and in the labour and union movement, socialists and anarchists were divided between pacifists and patriots; and within the regime, republicans were divided between radicals and moderates, Francophiles and Anglophiles, interventionists and non-interventionists. The proliferation and aggravation of divisions, together with the intersection of domestic policy and foreign policy dictated by the war, had immediate consequences on Portuguese politics. Domestically, an escalation in political violence and governmental instability took on huge proportions. Yet domestic policy had another, possibly even more important, effect on foreign and defence policy. The internal divisions that tore the country made national consensus impossible as regarded Portugal's stance vis-à-vis the war, and, in the absence of such a consensus, political objectives and national strategies shifted according to the political forces in power: the attempt at equidistant neutrality and non-intervention during Pimenta de Castro's regime; undeclared and Anglophile neutrality, with military intervention in the African theatre, during the moderate coalition governments; and the search for active belligerency and intervention in the European theatre during the radical governments of the Democratic Party. But although it is a fact that domestic policy has an impact on foreign policy, the reverse is also true. And, in fact, political change in Portugal, that is, the endless downfalls and replacements in government during this period are mostly driven by external motivations, that is, they aim precisely at a change in Portugal's position regarding the war: a change in objectives and strategies. Sidonism would be its most eloquent instance.

This third explanatory thesis for Portugal's entry in the European war means, first and foremost, to ponder over the domestic front, and tackle the subject on a different assumption:[12] in the first place, envisaging the entry into war as the outcome of a specific goal and strategy, namely the radical and interventionist strategy of the Democratic Party; secondly, beyond the consideration of both external and internal factors, the specific mechanism of their mutual interaction and the political exploitation of external factors and of the international conjuncture to pursue internal objectives.

There is no doubt that the interventionist strategy incorporated external objectives and took on the defence of national interests: the guarantee of colonial integrity in Africa and national sovereignty vis à-vis Spain, as well as the attainment of international prestige by the regime were indisputable objectives and echoed national interest. The radical and interventionist strategy was not only contingent on, but also actively pursued, internal objectives. Indeed, it took advantage of the international conjuncture offered by the war to accomplish its domestic policy goals. In the face of the divisions within and outside the regime, only an external threat and a large scale military intervention in the war could warrant the sacrifice of all the internal divisions and factions in the name of national interest and unity. Only that superior objective could conjure the political unity of all republicans, and perhaps even all the Portuguese, around a "national government". The party system of the Republic being a dominant-party system, it is evident that the "national mission" could only rest with that very party: the Democratic Party. It was not by chance that it emerged as the great champion for the radical strategy that supported Portugal's entry in the European war. It was not for nothing either that the party's official discourse invariably and univocally stated that the Republic's external and internal defence required an involvement in the war. At the head of a national government that managed to achieve political unity among republicans and the union of the Portuguese people, the democrats hoped to be able at once to neutralise extra-parliamentary oppositions and to engage the moderates in their political project. In this fashion, they sought to achieve the political and governmental stability so sorely needed by the Republic. They endeavoured to manage, moreover, to pass off as national that which was in fact their own party project. Their initiative resulted in the reinforcement of democratic hegemony within the party system and, simultaneously, the political consolidation of the republican regime. One final and never accomplished objective remained: the symbolic legitimacy of the Republic. Portugal's military intervention in the European war theatre was meant to ensure exactly that. In the national imaginary, Portuguese soldiers would fight in a key war theatre, alongside the great and amidst the great, under the flag of the Republic. Battles fought in the Great War would now be registered in the long story of Portuguese military feats as the last chapter of its heroic memory. The Republic was irrevocably enrolling in the gallery of history and national political imaginary, and that heroic memory finally awarded the regime, in retrospect, the legitimacy it had been lacking. Portugal's entry in the European war theatre would presumably ensure at once the political consolidation and the symbolic legitimacy of the Republic. As to the party, it would ensure its longevity in power.

In joining the fight, Portugal pursued a variety of war goals. A number of them were explicitly mentioned in the government's official report[13] and sought to justify active belligerency and military intervention in the European theatre; others were tacit and undeclared, but no less important.

In short, Portugal cooperated discreetly and unofficially with England and intervened militarily in the African theatre of war to ensure colonial integrity. It cooperated overtly and officially with England and the Allies, and took on a position of belligerency to ensure national sovereignty and a place in the concert of nations. Lastly, it opted for active belligerency and full military intervention in the European theatre, not only to protect the above-mentioned interests, but more than that, to ensure the consolidation and legitimation of the Republic and the reinforcement of the party whose strategy led Portugal to the Great War.

The conduct of war: War policies, the military instrument and the operational theatres

Whereas the entry into war is an essentially political matter, the conduct of war is fundamentally a military one. And, from a military point of view, there were two problems to consider in Portugal's intervention in the conflict: firstly, the state of the military instrument, that is, the conditions of preparedness and the equipment of the Portuguese Armed Forces upon their deployment; secondly, the conduct of war itself, that is, the strategic objectives and the military operations in each of the theatres.

On the eve of the Great War, the state of the military instrument of the Portuguese Republic was not so great. Above all, it was inadequate, both in terms of technical-professional preparedness of the forces and of equipment and armament, for the type of conflict it was about to be involved in. It might have been suitable enough for the African front, but surely not for the European front, especially the trenches of Flanders.

Emerging victorious from the revolution of the 5th of October of 1910, the Republic made the military question one of its political priorities. During the revolution, the sentiment and the attitude of the vast majority of the officer corps regarding the monarchic regime had been one of estrangement.[14] And that very sentiment and attitude dictated the neutrality that guided the behaviour of the majority of military units during the progression of the revolutionary events. It was said that, at the military level, the revolution was no more than a duel between the only two commanders who, on their respective sides, showed political conviction and military decisiveness: Paiva Couceiro among monarchists and Machado Santos among republicans. But one thing the republicans knew well: even if the Armed Forces had not defended the Monarchy, that did not mean they supported the Republic. Much to the contrary, as the republican regime became more radical, aloofness gave way to hostility.

Therefore, in the formulation of its defence policy and its military reforms, the Republic invariably addressed a double concern: the external – colonial and peninsular – threats weighing on the country, and, most importantly, the internal political question of the affirmation and consolidation of the regime.

In this context and driven by these objectives, the Republic endeavoured to undertake an in-depth military reform in the pursuance of a double objective: on the one hand, the modernisation of the Armed Forces according to the democratic idea of the "nation in arms" – a militia army; on the other, and at the same time, a politically reliable military instrument, that is, one that did not jeopardise but instead supported the consolidation of the republican regime.

The military reforms of 1911 extended to various fields. The first reform targets the model of military recruitment and is soon materialised in the new Law of Military Recruitment published in March of 1911. The goal was to replace the permanent army that was the Monarchy's with a Swiss-like militia army, perfect and democratic model of the "nation in arms". Universal and compulsory military service was instituted, as well as the system of militia officers. The first preparatory training was carried out at 17 years old and incorporation at 20, while time spent in the ranks was reduced to the required minimum. According to the law, 15 to 30 weeks depending on the weapon or service. Once demobilised and during peacetime, citizens would annually attend, for the course of seven to nine years, a two-week "repetition school".

The second field of military reform is that of territorial organic structure, which reorganises the composition and geographic distribution of the metropolitan army's divisions, inspired by General Pereira Bastos and published by decree in May of 1911. The six military divisions inherited from the Monarchy now became eight, in mainland Portugal – Lisbon, Viseu, Porto, Évora, Coimbra, Vila Real, Tomar and Braga – to which were added the Azores and Madeira commands. Each of the military divisions comprised four three-battalion regiments (regiments of infantry, cavalry, reserve squadrons and machine gun groups). The military personnel were supposed to be integrated by militia officers trained in "cadre schools" for a period of eight weeks.

A third field wherein Republic reforms had an impact was that of the separation, in the scope of internal security, between national defence against external threats and the enforcement of public order. The jurisdiction of the Municipal Guard, during the Monarchy, was limited to urban areas, whereas the Army was assigned with rural patrol. The creation of the Republican National Guard as a special police corps, aimed at rural policing, establishes a clear separation between the function of the military and that of the police. Although the creation of the Republic National Guard was motivated by military and security reasons, naturally there were also political concerns involved, namely the institution of a military instrument on which the republican regime might politically rely.

Finally, a fourth, last and no less important field of reform was that of equipment and armament. Throughout the second half of the 19th century and the first years of the 20th, the most significant technological advancement in armament had taken place in the artillery sector, namely, the replacement of bronze with steel in weapon manufacture, which granted weapons increased heat and impact resistance hence making it possible to increase the quantity of explosive material. Krupp's workshops, in Bismark's Germany, were pioneers in this process. Portugal, having attained a good level of proficiency in the manufacture and use of bronze in the Army Arsenal, during the first half of the 19th century, now finds itself technically outdated and unable to keep up with technological innovations in the field of armament, at once due to its lack of raw materials and know-how in the manufacture and use of steel. From that point onwards, Portuguese military industries retract from cutting-edge technology and shift exclusively to equipment and light armament (uniforms, harnesses and ammunition), leaving Portugal totally dependent on the importation of heavy weaponry.

In 1874, Portugal had already received its first shipment of steel material from Germany, and there are records accounting for the importation, between 1904 and 1906, of a set of 7.5 cm Schneider cannons from France, state-of-the-art at the time, which would in fact be requested by France during the 1914–1918 and would set off one of the first episodes of Portuguese participation in the conflict. The Republic tried to keep up with the most relevant innovations even prior to the war, creating a Heavy Machine Gun School, a military car park and taking its first steps in Military Aeronautics. However, when it came to heavy weaponry, necessitating cutting-edge technology industries, investments fell radically short – and well below those made during the Monarchy – of the requirements of military modernisation.

The military reforms of 1911 had been designed to adjust the Armed Forces to the Republic, and were supposed to be in-depth and long-term reforms. They had a short life, however, and yielded poor results. In force for the duration of two years, and meeting with little success, they were suspended in 1914, when the war broke out. Some had not even been implemented.

Compulsory and universal military service was undermined by the old practices and the preservation of the bureaucratic apparatus inherited from the Monarchy. The rate of draft evaders remained high and the number of citizens that actually performed military service remained low. In 1915, for instance, of the 70,000 called up only 22,862, that is 32.9%, actually reported.[15] Of those having undergone assessment, many were discharged in accordance to the old rules, although now without the benefits of "remissions", that is, pecuniary compensations that freed the drafted from the obligation to do service. This system was kept in place between 1911 and 1914, having been briefly and partly suspended during the war and then again resumed between 1919 and 1925.

On the other hand, the model of the militia army found itself burdened by the maintenance of a relatively large corps in the permanent cadre: 1,773 officers and 9,926 privates.[16] This corps continued to prove necessary for two compelling reasons. In the first place, the requirements of maintaining the pacification campaigns that went on in the overseas territories: in Angola between 1911 and 1914, in Mozambique between 1912 and 1914, as well as in Guinea, India and Timor in 1912. The second reason was of a political nature and had to do with the situation in the peninsula: the monarchist incursions of 1911 and 1912 had reinforced the idea of their necessity for the military defence of the Republic.

Similarly, the civic function of the militia army and the training of the citizen-soldier were suspended. The so-called "repetition schools" would be closed for more than two years, between 1912 and 1914.

Likewise, the separation between national defence and internal security was clearly not observed. The Army was often called, between 1912 and 1914, to intervene in issues of public order, particularly strikes and urban social conflicts. And the Republic National Guard never ceased to be regarded as a second military force deserving of the political trust of the Republic.

As we have seen, when it came to armament, the balance was definitely negative. And this was no less true for the Navy than for the Army. An ambitious plan for naval rearmament, inspired by the young Second Lieutenant Pereira da Silva and approved by the Parliament in 1912, was never implemented due to the State's noticeable lack of financial capacity.[17] The outbreak of war in 1914 called a halt to what little remained of the 1911 reforms.

The Republic did not meet with greater success in civil-military relations. It had believed that an army of citizen-soldiers and a corps of militia officers trained in the reformed Army School would ensure the liaison between the military and political power. Instead, it was faced with the failure of the militia model and the preservation of a conservative and progressively hostile to the Republic corps of officers. Deteriorating social-professional issues, the impossibility to commit to rearmament and the political evolution of the regime itself did the rest.

In its ambitious military reforms, the Republic pursued two chief objectives: the modernisation of the Armed Forces and the establishment of a military instrument favourable to the regime. It managed to do neither.

Yet those were precisely the two essential conditions required to engage in war, particularly on the European front, which did not gather political consensus. The Republic was lacking, but urgently sought to assemble a well-prepared and equipped military instrument and an elite corps in synch the government's war goals.

If, however, Portugal's diplomatic position was one of feebleness before the war, the condition of the military instrument at its disposal to handle the conflict was no sturdier.

Portugal would fight on three battle fronts and two theatres of war. Initially, during the years of neutrality, between 1914 and 1916, in the African theatre, on two fronts: Angola and Mozambique. Later on, after entering the war, between 1917 and 1918, in the European theatre on the Western front, in Flanders.

At the colonies, even if not *de jure*, Portugal was *de facto* at war since mid-1914. It all took place, however, in a secondary and peripheral theatre, which did not require a declaration of war or even belligerency.

In August of 1914, in Maziua on the north of Mozambique, Portuguese troops were involved in the first military incidents with German troops. And in October of the same year, the Germans would attack the forts of Naulila and Quangar. The campaign went on, with a German incursion through the Angolan hinterlands. This incursion was accompanied by an attempt to incite indigenous populations to rebel against Portuguese rule. These first confrontations reached their peak on December of 1914, with the Battle of Naulila.

They opened the two war fronts that Portugal would face in the colonial theatre, whose unanimous war goals were: in the first place, an essential condition, the preservation of the territorial integrity of the African empire; and secondly, if at all possible, the territorial expansion of the empire and its spheres of influence, at the expense of either the German colonies or the subjugation of indigenous populations.

Following these first military incidents in Africa, and according to the outlined objectives and the general agreement among political leaders and the public opinion on the need to preserve Portuguese sovereignty over the colonial empire, the first contingents of troops were deployed to the African theatre. Two mixed detachments (infantry, cavalry and artillery troops), each comprising 1,500 men, were then deployed. The first, commanded by Alves Roçadas, embarked for Angola, and the second, commanded by Massano de Amorim, for Mozambique.[18] Military confrontations in Angola were more severe than in Mozambique, and led the Portuguese government to decide for a reinforcement of the garrison by sending an expedition led by Alves Roçadas: another 2,400 men in November and 4,300 in December.

As is always the case with war scenarios, numbers are controversial and, both at the time and later, at the hands of historiography, give rise to disparate estimates. However, *grosso modo*, regarding the deployment of Portuguese troops to Africa, overall figures do not seem to be far from the estimate presented by the Portuguese government at the Versailles Peace Conference: around 34,600 metropolitan Armed Forces fighters joined by around 19,500 indigenous soldiers.[19]

Despite their distance from the European theatre in which the fates of war were played, the Portuguese troops continued their operations on the two fronts of the African theatre – southern Angola and northern Mozambique – prior to and after the declaration of war and Portugal's entry in the European theatre.

In Angola, under the command of General Pereira d'Eça, since March of 1915 the Portuguese troops are headed towards a major strategic goal: the occupation of the Lower Cunene. In this southern Angola campaign, the reoccupation of the Humbe and the Battle of Mongua would represent the most relevant military watersheds.[20]

In Mozambique, military strategic goals were essentially two: the reoccupation of Quionga, a Portuguese territory occupied by the Germans since the late 19th century; and the passage of the Rovuma River and the occupation of a southern portion of the the German colonial territory. The fighters of Maúta, Nevala, Negomano and Nhamacurra would stand out as the more relevant military watersheds in Mozambique's campaign.[21] As we shall see, despite all the hardships involved in fully achieving the objectives set for the African theatre, the outcome of the operations and the overall upshot of the war enabled Portugal to fulfil its colonial objectives: the preservation of the Empire's territorial integrity and the reclaiming of control over Quionga.[22]

Although gathering the consensus of Portuguese society, for serving that nationwide objective which was colonial sovereignty and the territorial integrity of the empire, the military intervention in the African theatre was not decisive. Neither diplomatically nor militarily. The epicentre of the war was unfolding in Europe. For that very reason, the big question, both the most controversial and the most important question faced by Portugal, was that of a possible intervention in the European war.

More than in the colonial theatre, intervention in the European theatre required the fulfilment of two major political-military conditions: first, a well-prepared and equipped military instrument capable of tackling the new type of conflict that was trench warfare, of which the Portuguese Armed Forces had no experience and for which they had no training; second, an elite corps faithful to the government and in tune with the political objective of European interventionism led by the more radical republicanism of the Democratic Party. The integrity of the colonial territories and the intervention in the African theatre were not only supported by the Portuguese civil society, but also by the corps of officers which felt militarily capable and used to the pacification campaigns in Africa. Now, this was not the case at all with the European theatre. Deciding for an intervention in the European war thus opened deep fault lines within Portuguese society, and a good share of the corps of officers, predominantly conservative, could not identify with the political objective of intervening in a European matter, to which a technical reason was added and would forever be invoked: the lack of military preparedness for the new type of warfare that was trench warfare.

This is how the majority of the officers of the permanent cadre marched towards the African campaigns, whereas to the trenches of Flanders would head mainly the militia officers led by a team of "young Turks" commanded by Norton de Matos. Close to the Democratic Party, it would soon be dubbed "the war team" and lead the whole military intervention and the war policy itself.

During the course of the war, the base of military recruitment is broadened and universal and compulsory military service is in fact enforced. At the same time, the schools for militia officers open their doors and the future Expeditionary Corps begins to take shape.

In truth, since the victory of the interventionist forces after the revolution of the 14th of May of 1915, a war policy favourable to the creation of the so-called "Auxiliary Division" – envisioned since the onset of the war to intervene in the European theatre – began to develop. That division was given the name of "Training Division" and plans were made for the training camp that would later occupy the future Tancos military polygon, known as "the city of *paulona*" (*pau* and *lona* meaning pole and canvas, in a reference to the swarm of tents which were used to quarter the soldiers).

Although the idea had been entertained since May of 1915, only in February of 1916 does it take shape and the project is implemented, on the verge of Portugal's entry in the war and following the English requisition of the German ships stationed in Portuguese ports, which would lead to Germany's declaration of war in March of 1916.[23]

Faced with substantial difficulties having to do with insufficient training, scarce availability of horses and mules, and lack of armament and ammunition, the first troops conclude their period of intensive training in August of 1916 and stand ready for the European war. For some, it represented the "miracle of Tancos"; for others, a prelude of disaster.

Between August and December of 1916, Portugal, France and England engage in diplomatic-military negotiations over the terms of the cooperation between the Portuguese Expeditionary Corps (CEP) and the British Army: the obligations of each party, organic structure, tactics and even the armament used in modern warfare.

In January of 1917, a Convention is signed regulating the employment of Portuguese forces in Britain's operational area.

The Convention stipulated that the Portuguese expeditionary force would be composed of 35,000 men. Soon after, however, in February, by proposal of the CEP's commander – keenly supported by the political power and previously approved by England – the number would rise to 55,000 men divided into two divisions.

Militarily, this decision had profound consequences: the two divisions would make up an army corps. Politically, this was invested with tremendous significance: should the war end before the end of 1917, this was a clear and unambiguous sign of the total commitment of Portuguese interventionism.

Of the total number of deployed soldiers, the large majority – around 65% – belonged to the infantry; in second place, 17.5% belonged to the artillery; engineers came third with 7.5% and, finally, cavalry, representing a mere 4%. The remaining personnel belonged to Health Services and Military Administration.

Around 28% of the total mobilised troops were volunteers and, except for a few exceptions, the remaining 72% represented the great mass of the recruited. Moreover, the ratio of officers to soldiers was 1 to 15.3.[24]

For all the impressiveness of the total number of mobilised forces, leaves of absence, disease and, from the late 1917, difficulties in the *roulement* meant they were never fully operational.

According to the convention of the 3rd of January, the transportation of troops was to be carried out by sea under England's supervision. The agreement would never be fully observed. Between February and September of 1917, the transportation of the CEP troops was shared between 38 British ships and eight Portuguese ships. From that moment onwards, the shipping shortage, aggravated by submarine warfare and, it was said, widespread fear of contracting the Spanish influenza, would keep England from fulfilling the deal and require that Portugal ensure the transportation of troops on its own, with two ships making four distinct journeys. From the debarkation port to the concentration zone Aire-sur-la-Lys – Portuguese troops were transported by railway.

The sector of which the CEP would be in charge was located in the south of Flanders, precisely in the valley of the river Lys, from between Armentières to La Bassée and Merville to Béthune. It stretched across a long front, which in the course of the war varied between a maximum length of 11 km and a minimum of 4 km, according to the evolution of the military campaign.

Before taking over responsibility for its sector, and in compliance with the convention signed on the 3rd of January of 1917, the CEP troops had to undergo a period of intensive training mentored by the English Army. Subsequently, still within the scope of the training, the troops would gradually and one step at a time experience the daily routine of the trenches. The next phase would be taking on responsibility for defence, but under the control and incorporated by a subsector of the British brigade. Finally, when all the troops had gone through all the stages, the traineeship would be over and they would be ready to assume full responsibility for their sector. That is, only when the three Portuguese brigades had successfully completed the various phases of the training programme would the Portuguese division be deemed capable of autonomously defending the front.

This would be the case when, in July of 1917, the three brigades of the 1st Division took charge of a sector under the command of the XI Corps of the British Army. And it is only in November of 1917 that the CEP commander, General Tamagnini de Abreu e Silva, takes full charge of the Portuguese sector.

The Portuguese front, as all fronts in trench warfare, was composed of three lines of defence, and troops were distributed according to those three different yet complementary spaces:

- one first line of defence, next to "no-man's land", was composed of two trench lines: the front trench – line A – was protected by barbed-wire entanglement, and another one, 300 to 880 metres to the rear – line B – was a robust and continuous line, also protected by wire mesh and sided by the heavy artillery; between these two lines were located the support trenches.
- behind the first line was the intermediate line, also known as "village line": around 3,000 to the rear of line A, it was composed of a mixture of ramshackle villages and campaign fortifications, and constituted in fact a reserve or back-up trench line.
- finally, the second line, or body line: 6,000 metres from line A, usually it was heavily fortified, and connected to communication trenches to the west.

The "no-man's land" was the space between the two enemy front lines, between 80 and 250 metres apart. It was a mythical place of death and heroism.

It was in this operational theatre that, from February of 1917 to April of 1918, between artillery bombardments and assaults to the enemy line, the trench war unfolded for the CEP troops.[25]

During this period, Portuguese defence lines were the target of 63 offensives by German troops and around 20 bombardments. They retaliated with the artillery at hand and launched around a dozen assaults against German troops, unsuccessfully attempting, on eight occasions, to break through enemy lines.

In these 13 months of trench warfare, the imperial army had unarguably the upper hand: according to the available and always controversial numbers, the Germans suffered around 123 casualties (eight dead; 56 wounded; 59 prisoners), against 627 on the Portuguese side (107 dead; 358 wounded; 162 prisoners).

The drama of human suffering notwithstanding, and regardless of the accuracy of the estimates, none of these numbers is truly enlightening, either from the military or the political point of view. For Portugal, its decisive battle in the context of the First World War would only come on the 9th of April of 1918: the Battle of the Lys.

In the interim, however, between the winter of 1917 and the spring of 1918, material conditions and the morale of the Portuguese troops had gradually but swiftly deteriorated. And there was nothing that the government or the military command in Lisbon could do to rotate the troops.

Now, this would be, for Portugal, the central issue of the conduct of war, which would become known in the Portuguese history of the Great War as the *roulement* problem, that is, the relief in place of the CEP contingent.

Since September of 1917 the shortage in British tonnage had halted the sea transportation of Portuguese troops to the war front, which, as we have seen, became totally dependent on the reduced capacity of the Portuguese Navy.

On the other hand, ever since Sidónio Pais's rise to power, the government – without altering the diplomatic status of the country – launched a new direction for war policy, especially regarding the military. In January of 1918, a new convention with England is signed, changing the composition of the CEP: the two original divisions would be reduced to one, which would become tactically dependent on British command.

If this alteration possessed military relevance, its political significance was even greater. It was the sure sign of a change in Portugal's attitude towards the war: it might not have meant an outright break, but it undoubtedly implied a shortfall in Portugal's commitment to its allies.

For the troops, however, graver than the decrease in numbers was the question of the *roulement*: since late 1917, there was no more relief in place of the contingent and no more reinforcements were forthcoming.[26]

In public debate at the time, as in subsequent historiography, the question of allocating responsibility for such a conduct in war was discussed. For some, Sidónio Pais's war policy was totally accountable; for others, England's breach of contract and shipping withdrawal was responsible. Perhaps it would be fairer to acknowledge a shared responsibility.

Far from the public debate, the consequences were dramatically felt in the battlefield. Growing physical strain caused by long months in the trenches, refusal of leaves due to transport shortage, the rigours of winter, the rising fierceness and frequency of enemy assaults and the lack of reinforcements steadily worsened the morale of Portuguese troops. Insubordination and desertion begin to spread. Between 1917 and 1918, a total of 372 soldiers were condemned in the CEP military courts, of which only one, charged with treason, was sentenced to death. To crime, disease is added, and finally suicide: 10 recorded cases.[27]

During March of 1918, German attacks escalated. On the 6th of April, the convention of the 21st of January of 1918 is implemented and the 2nd Division of the CEP came under the tactical command of the British Army. The front, theretofore divided into four sectors, is reduced to three, with no change in its reach. On the eve of the great battle, the dreadful state of the CEP was so noticeable that the British command decides to pull the Portuguese division out of the front line and keep it in a rear position, as reserve to its army corps. The order issued on the 8th of April had scheduled the replacement for the 9th.

Too late. It was 4:15 in the morning of the 9th of April when a gigantic bombardment broke out along the Portuguese front. The Battle of the Lys had begun.[28] The military outcome is well known: the division was overrun and the allied troops were forced to retreat.

Altogether, notwithstanding exercising caution in the numbers, it is possible to risk an estimate of the CEP casualties: 1,341 dead; 4,626 wounded; 1,932 missing; 7,740 prisoners.[29]

This was Portugal's great defeat. After the battle, nothing would remain the same, either militarily or politically.

Militarily, three infantry battalions were still formed with the remnants of a crushed CEP. Incorporated in the British Army, they fought on the front lines until victory and the Armistice. Politically, however, the toll was heavy and became apparent in the course of the Peace Conference.

Managing peace: The aftermath of the war and the Versailles Conference

"Did Portugal lose the war?", the headline of a popular newspaper asked on the 9th of May of 1919. What a bizarre thing to ask of a country that had been involved in conflict with Germany, in Africa, since 1914, and that had officially entered the war alongside the Allies in March of 1916. Of a country that had fought on two fronts in the African theatre and on the Western front of the European theatre. That had paraded in the victory celebrations, under the Arc de Triomphe in Paris, next to the Allies and in the capacity of victorious power. There was no doubt about it. But still the feeling of misgiving became lodged in the political milieus and in public opinion. And the misgiving was legitimate.

By the end of the war, Portugal was worse off than in it had been at the start. A torn economy, social frictions and political instability. On the other hand, the Peace Conference was experienced in Portugal with a bitter feeling of injustice. Portugal did not find compensation in peacetime for all the effort it had put into war. Granted, it stood among the victors, but victories are not all the same. There are contented victors and dissatisfied victors. Victors who achieve their war objectives and victors who fall short of them. And, truth be told, the appraisal of victory resides in that comparison between the objectives set out when entering the war and those attained by means of peace treaties.

Accounts from the period[30] and Portuguese historiography[31] have undertaken this exercise for post-First World War Portugal. The preliminary, clear and explicit definition of war objectives by belligerent governments is always a delicate – not to say hazardous – exercise. The clarification of specific objectives for entering into war may carry two types of risk: in the first place, by making them public at the beginning of the conflict, the government may be exposing itself to a sentence of defeat in case these expectations fail to be met by the end of the war; secondly, by actually listing tangible – either economic, political or territorial – objectives, the government may compromise the heroic character of the war effort, which usually finds sufficient validation in principles and values. This being said, and although among political leaders and military staffs there is invariably a more or less clear definition of the war objectives, the majority of governments have preferred, when entering a war, to declare vague intents, largely wrapped up in a rhetoric of heroism. This way they spared themselves being judged as the result of a future balance between the sacrifices endured and the goals achieved, and could focus their attention on the war effort and victory itself, ensuring at the same time the soldiers' morale, the mobilisation of civil society and the legitimisation of war. In the Portuguese case, these two factors were joined by a third, which resulted from the violent controversy between interventionists and non-interventionists: the absence of a nationwide consensus over the very purpose of war.

In January of 1917, when the first troops left for the European theatre, the Sacred Union government laid out the national goals. It called them "Clear Words – reasons for Portugal's military intervention in the European war".[32] More than a formulation of war goals, it offered arguments in favour of Portugal's involvement in the European war. It was addressed to every Portuguese, but especially to those who were setting off. It recalled, moreover, "the superior reasons that led the Homeland to ask for their sacrifice and their heroism". The rhetoric of heroism was necessary to boost the morale of the troops, but also to encourage a nationwide consensus, which it never managed to secure.

What reasons, then, led Portugal to become involved in the war? In the first place, this being a war of alliances, and Portugal being bound to England by the oldest of alliances, it could not remain neutral in the face of the state of war between England and Germany. In the second place, because Portugal, being non-belligerent until 1916, had been the target of several military attacks – a number of them directed at Portuguese merchant vessels – and encroachments on sovereignty in the colonial territories, namely southern Angola and northern Mozambique. Thirdly, the proximate reason that had prompted Germany's declaration of war: the request, from England, that Portugal seized the German ships stationed in Lisbon ports. To justify the requisition, the Portuguese government invoked the consequences of war, the state of the merchant navy fleet and the state of the country itself, which was in no position to relinquish those resources. Finally, the selfless and sympathetic nature of the Portuguese participation, that is, the rhetoric of heroism: "we are neither driven by conquering ambitions nor by a thirst for rewards. The superior interest that guides us (...) is that of further strengthening our alliance with the noble English nation (...), reinforcing it with our own efforts and sacrifices, treasuring and exalting it, which results in us exalting and treasuring ourselves". In fact, two nationwide goals prompted Portugal's entry in the war and both were dependent on the Anglo-Portuguese Alliance. One pertained to the colonies and the other concerned Europe and the peninsula. Not to mention the economic and financial objectives, and those relating to domestic policy, that emerged further ahead. The colonial goal was clear-cut and well-defined: the preservation of the Empire's sovereignty and territorial integrity. It represented the chief national concern and the single agreed-upon objective. Agreement over this point ensued from the acutely perceived threat hanging over Angola and Mozambique, stemming both from Germany's colonial pretensions and Britain's acquiescence, which had indeed been corroborated by the Anglo-German agreements in 1898 and 1912–1913 over the partition of Portuguese colonies. Entering the war side-by-side with the Allies and under the blessings of the Anglo-Portuguese Alliance was the most effective insurance against the German threat, but also, at the same time, the best way to strengthen the alliance and to limit England's leeway on the negotiating table by the end of the conflict. The European and peninsular goal stemmed from the fragile international standing of the Republic. From Portugal's vulnerability in the face of the diplomatic rapprochement between Spain and England, since the Cartagena Pact in 1907, but especially the annexationist pretensions of the Spanish Monarchy and England's leniency towards Spain. That posture was made clear on two occasions prior to the war: the monarchist incursions of 1911 and 1912. Hence, by entering the war beside England and under the aegis of the Anglo-Portuguese Alliance, Portugal manage to achieve a two-fold objective: first, the diversification of its international position. Vis-à-vis neutral Spain, a belligerent Portuguese reinforcing its alliance with England. Secondly, through this, it strove to earn the much coveted place in the "concert of nations". The first objective, which gathered consensus, might have been achieved through the preservation of neutrality and the concentration of military efforts in the colonies. The second, on the contrary, divided Portuguese society and necessarily entailed belligerence. Domestic objectives aside, this latter aspiration was the decisive motivation for Portugal's decision to enter the war. It was also what earned it a seat at the League of Nations (LN).

By the time the Peace Conference was held, however, goals had changed. Two and a half years of belligerence and intervention in the European theatre had tremendously impacted the economic, social, political and military levels. The country was not the same and neither were its objectives.

The Portuguese delegation to the Peace Conference underwent a change of leadership as a consequence of the internal political change brought about by the fall of Sidonism and the restoration of the "New

Old Republic". A change that concerned, moreover, the controversy over the conduct of war, which replicated and updated, under a new avatar, the old contention over Portugal's intervention in the war. The two presidencies, Egas Moniz's and Afonso Costa's, echoed two different outlooks on what were the national goals. The political positions they represented had always held contrasting views on the subject of Portugal's entry and conduct in war. What were now, in a context of peace management, Portugal's ambitions at the Versailles Peace Conference?

The colonial factor, which had been a major motivation for entering the war, did not represent a genuine peace goal. The territorial integrity of the empire derived directly from Portuguese belligerence. Portugal's claim to Kionga, a Portuguese territory in northern Mozambique occupied by the Germans in the late 19th century, was upheld. However, the Portuguese delegation regards the recovery of this territory not as reparation, but rather as the restitution of a possession, which was tantamount to reinstating a right. On the other hand, ruled out as it was any South African claims over southern Mozambique or Belgium's over the Cabinda enclave, Portugal does not request any other territorial compensation in Africa. It would not, unless Belgium came to be granted a parcel of any of the German colonies, which did not happen. Portugal was pleased with regard to the colonies and, in the main, its war goal was achieved: the territorial integrity of the African empire.

However, Portugal's peace goals differed from its war goals. And in the aftermath of the war, the rising political tribulations and the urgent need for economic reconstruction had significantly altered priorities.

Diplomatic efforts, which had been common to both presidencies of the Portuguese delegation, now focussed on economic and financial conditions and were directed towards four major objectives: circumvention of the payment of war debts; the right to reparations and indemnities from Germany; participation in the reallocation of the German ships requisitioned by England in 1916 and the constitution of a merchant fleet; and, finally, the attempt to seize peace negotiations to obtain military hardware from the defeated powers and thus reequip the Portuguese Armed Forces: Army, Navy and Air Force. To a greater or lesser degree, the economic objectives were almost all fulfilled.

The same cannot be said of the big political objective. Apart from the colonial question and the economic concerns, there remained still the key political issue handed down from the second war goal: the conquest of "a place in the concert of nations". What had been a vague and abstract formulation before the war now became concrete and well-defined: Portugal's active presence in the process of international restructuring in the aftermath of the war and, in particular, its participation in the Executive Council of the newly-created League of Nations (LN).

As had happened before the war, this once again emerged as the foremost source of dissent between the two presidencies. Unlike Egas Moniz, who does not pursue that kind of political goal, Afonso Costa directs the bulk of his diplomatic efforts to securing a seat for Portugal in the executive Council of the League of Nations (LN). What might have been a great political victory never came to happen. On the contrary, it would represent the great failure of national goals. An absolute failure, since the Portuguese candidacy was not even accepted. An even graver failure in relative terms, because neutral Spain would manage to attain what belligerent Portugal could not. This meant that Portugal was unable to obtain the much yearned for place in the "concert of nations".

Now, at this point, it seems useful to ask, as the press did at the time: Did Portugal lose the war? Yes and no. It did not, in the sense that it was a belligerent country that had fought beside the Allies and taken its seat as a victorious power at the Peace Conference. Moreover, because it managed to fully secure its colonial objectives and, to an extent, also its economic objectives. But it also did lose it, since it failed to achieve, in the peace treaty, its primary political goal, that which had ultimately driven it to war. The thwarted international aspiration, together with the never fulfilled domestic objectives, was the unarguable sign of a negative balance.

By the end of the war, it was clear: the economic and financial situation, social instability, party disaggregation and the crisis of institutions had not only hampered the implementation of the Republic's modernising project, but also, in the end, precluded the regime's political consolidation and nationwide legitimation. Moreover, it had opened the door to the authoritarian drift that largely emerged from the rubble of the First War. All in all, Portugal came out victorious. Yet, in fact, victory tasted bitterly like defeat.

CHAPTER

4

The *Estado Novo* and the Spanish Civil War

The international standing and the *Estado Novo* before the War

On the 28th of May of 1926, the military dictatorship overturns the republican regime and puts an end to sixteen years of democratic Republic in Portugal. It did not mean, however, any major change, either in the international standing of the country or the premises of its foreign policy. In the throes of a serious financial crisis and engulfed by its internal political contradictions, the military dictatorship stuck to following the external orientations of the First Republic: the colonial project, the Anglo-Portuguese Alliance and even the commitment to multilateral diplomacy within the League of Nations (LN). No significant changes to the country's international standing would take place until the 1930s, with a shift in the international conjuncture and the institutionalisation of Salazar's authoritarian regime.

In 1933, the year marking the establishment of the *Estado Novo*, Salazar voiced, in his first interview, his so far unclear ideas on Portugal's international standing and foreign policy. Basically, he reacts against Geneva's "international parliamentarism" and reaffirms the precedence of the Anglo-Portuguese Alliance. It all pointed to a return to Portugal's traditional principles guiding external affairs, which actually proved to be the case. In his first foreign policy statement, in 1935, he corroborated that trend and outlined very clearly the main thrust of his foreign policy: in the first place, overt criticism of the League of Nation's parliamentarism, which he regards as the centre of continental politics, restating, on the contrary, Portugal's Atlantic vocation and strategic detachment from central European affairs; second, the reaffirmation of the Anglo-Portuguese alliance as the central axis of his foreign policy; third, reinforcement of the peninsular friendship which he deemed crucial to the stability and political longevity of his regime; and finally, the defence *à outrance* of colonial integrity.

Hence, on the eve of the Spanish Civil war, Salazar had already clearly outlined the strategic options that would steer Portuguese foreign policy in the two dramatic international events that were forthcoming: the Spanish Civil War and the Second World War. Moments as internationally decisive for the country as they were, politically, for the regime. To such a degree that Salazar would decide to personally undertake the conduct of foreign policy and take up the office of Foreign Affairs, in addition to the Presidency of the Council, from the onset of the Spanish Civil War and until the aftermath of the Second World War – eleven long years, from 1936 to 1947.

In fact, detachment from the European question and the affirmation of an Atlantic Portugal and the African imperial project would make up Salazar's strategic priorities, as well as a return to bilateral diplomacy and to the historical balance of the Lisbon–London–Madrid triangle would represent the major concerns of his foreign policy in the course of the Spanish Civil War and the Second World War.[1]

The integrity of the empire having been secured and the colonial question settled since the aftermath of the First World War, throughout all of the 1930s, the country's chief international vulnerable spot shifts to the peninsula. In fact, from 1931 to 1939, the main concern of the *Estado Novo*'s foreign policy was completely tied up with the "Spanish question".

Within the Lisbon–London–Madrid diplomatic triangle, alongside the Anglo-Portuguese Alliance, the core issue was the so-called "peninsular friendship". Now, Salazar nurtured a steadfast and peculiar interpretation of this "peninsular friendship": it was a desirable principle and, moreover, an important factor of peninsular balance, but it was heavily constrained by its objective conditions, that is, by the political nature of the regimes. In such a small geopolitical space as the Iberian Peninsula, divided into two states, the difference in strategic potential, together with the political difference between the regimes, would amount to a twofold threat for Portugal: a threat to the State, in the first place – to national independence; and, in the second place, a threat to the regime – to the longevity and ultimately the survival of the *Estado Novo*. And in fact, between 1931 and 1936, the Spanish Republic would been seen by Salazar as a double menace: to the "Iberian threat", it had been added the "red threat".

Instituted in April of 1931, the Spanish Republic soon proved to be a support hub for the conspiratorial initiatives of the Portuguese "*reviralho*" (pro-insurrectional republican opposition to the dictatorial regime), who continued to foster republican resistance to the *Estado Novo*. But the Spanish Republic went even further during the republican-socialist government – the so-called "*bienio rojo*" (red biennium), between April of 1931 and September of 1933 – which confirmed the double threat and caused alarm in Salazar. Manuel Azaña, president of the Council and leading figure of Spanish republicanism, not only gave financial and political support, but also armed a number of those movements, especially the notorious "Buddhas". This support granted to opposition movements against the *Estado Novo* – an "Iberianist conspiracy"[2] – was indeed part of a strategy of "political modernisation" of the peninsula set about by democratic and republican Spain, which was supposed to lead to a federation of Iberian nationalities.[3]

In September of 1933, however, the rise to power, in Spain, of a moderate party, enabled a degree of normalisation and a period of appeasement in relations between the two peninsular states. On the Spanish side, initiatives – both in domestic and foreign policy – tending towards a diplomatic rapprochement are rife: from the opening of a parliamentary inquiry regarding the provision of weapons to the Buddhas in 1934, to the backing of Portuguese positions by Spanish delegates at the League of Nations (LN), when Portugal was head of the Sanctions Committee during the Abyssinia Crisis in 1935, to public endorsement of Portuguese colonial rights, when the dismemberment of the Portuguese Empire is once more the subject of discussion, to the proposal of a new trade agreement between the two countries.

During this short interregnum, if the Spanish peril was still alive, at least the revolutionary danger appeared to be fading. But not even that convinced Portugal to respond in like manner to Spain's display of cordiality, invariably regarded with misgiving. Portugal's reticence was justified by the victory of the Popular Front and the establishment of a left-wing coalition government, in February of 1936.

From that moment, the Spanish peril is revived, amplified by the revolutionary danger. It all validated and exacerbated Salazar's concerns regarding the political irreconcilability of the coexistence of two opposite regimes in the geopolitical space of the peninsula.[4] Salazar and the *Estado Novo* were thus faced with two distinct options: they could either collaborate with the Republic, maintain relations with the legitimate government and wager on the "Balkanisation" of Spain, which ultimately might result in the disintegration of the State and the emergence of a plethora of small states with differing political regimes; or, instead, support the nationalist insurgents and the establishment of a like-minded regime. Naturally, authoritarian. Salazar went for the second option.

From then onwards, tensions between Lisbon and Madrid escalate. In Spain, the insurrectionist activity of the "*reviralho*" emigrants is given a fresh impetus by the support of the Popular Front's government. In Portugal, the same thing occurs with the enemies of the Spanish Republic, who choose Lisbon as headquarters for their conspiratorial activities, with the acquiescence of the *Estado Novo*. Up until the summer of 1936, diplomatic incidents involving the two chancelleries multiply and, in July, ongoing negotiations for the bilateral agreement are ceased once and for all. Salazar was now certain that a connection between foreign policy and domestic policy was crucial. That the triumph of a left-wing government in Spain was inextricably linked to the revival of the "*reviralhismo*" in Portugal. And that, for the same reason, yet conversely, what profited the *Estado Novo*, both politically and in its bilateral relations, was the fall of the Republic and the triumph of nationalist forces in Spain.[5]

This conviction and the subsequent decision to support the nationalist forces of General Franco will lead Salazar to an important turning point in the country's external orientation, breaking up with one of the traditional principles of Portuguese foreign policy dating back to the very foundation of the regime of *Estado Novo*, in 1933. The principle, that is, of non-interference in Spain's internal affairs. All the more so because it carried a further risk of favouring the establishment, in the Iberian Peninsula, of a regime potentially hostile to England, with all the consequences it might entail for the Anglo-Portuguese Alliance.[6]

In short, on the eve of the Spanish Civil War, the international standing of the *Estado Novo* was clear: its major concern was the peninsular context, where the traditional "Spanish peril" was joined by the "red danger". For Salazar, it was not just about the external security of the State but, moreover, the survival and internal consolidation of the regime. For that reason, he decided to support the *alzamiento* against the Spanish Republic, which would require greater autonomy regarding England but strictly within the limits of the Anglo-Portuguese Alliance. A strenuous but not impossible exercise, as Portuguese diplomacy in the course of the Spanish Civil War would prove.

Pushing the limits of ambiguity: Portugal in the face of the Spanish Civil War

When, on the 18th of July of 1936, the coup against the Republic takes place in Spain, all had been arranged for a swift and decisive victory.

It did not work out that way. Soon the republicans reorganised their forces, armed their civil supporters and took control of the main cities. What should have been a fulminating coup turned into a prolonged and bloody civil war. Spain was divided into two opposing political-military camps and external support rapidly converted the conflict into an international issue, a "European civil war fought within Spanish territory".

In that context of civil war in neighbouring Spain, what will be the position adopted by Salazar and the *Estado Novo*? According to the strategic options and diplomatic principles previously laid out, Portuguese foreign policy will be aimed, during the Spanish Civil War, at the harmonisation of two seemingly conflicting imperatives: on the one hand, observance of the Anglo-Portuguese alliance, an instrument recognised as crucial for securing the African colonial empire in a Europe increasingly ruled by the expansionism of budding totalitarian regimes; on the other, support to the Spanish nationalists, which was deemed paramount to the establishment in Madrid of a regime that did not threaten the political survival of the *Estado Novo*. In order to conciliate the irreconcilable, Salazar's diplomatic strategy will remain ambiguous, in an effort to avoid at all costs the need to make a choice between the two imperatives. Granted, his strategy pushed the limits of ambiguity, but never infringing on the Anglo-Portuguese Alliance.

Hence, formally speaking, it abided by the Alliance and coordinated all his positions with the Foreign Office: it agreed to the policy of non-intervention, took part in the London Committee, complied with the inspection of the Portuguese borders and did not even recognise officially the Burgos regime without England taking the lead.

In practice, however, the opposite was true. It dealt prodigally with General Franco and spared no efforts in supporting nationalist troops: political and diplomatic support, financial and logistic concessions within Portuguese territory and even military support in the mobilisation of volunteers.

When, in the face of the internationalisation of the conflict, making use of a policy that already heralded the "appeasement", France and England proposed the idea of non-intervention in the Spanish Civil War, Portugal's accession was deemed essential. It was only natural, given the length and permeability of the Portuguese–Spanish border. Hence, London and Paris called for Portuguese participation. Salazar responded with dilatory manoeuvres and the imposition of conditions. It protracted its accession until the Committee was in operation, on the 21st of August of 1936. Precisely the time it took to hand out logistic support to the nationalist troops that, a day earlier, with the occupation of Merida and Badajoz, had made important strategic progress in the direction of Madrid. On the other hand, Salazar accepted the non-intervention pact on the condition that all defensive action against a potential "subversive regime" that might be established in Spain was not deemed an interference. This was his way of taking a stand and striving for political and diplomatic leeway, as well as a degree of autonomy, vis-à-vis England.[7] To a certain extent, his efforts were successful.

Both at the London Committee and at the League of Nations (LN), the stance of the Portuguese representatives never waned in its endorsement of the nationalist cause, from the objection to the internal legitimacy of the republican regime to the international backing of the *alzamiento*, to the external validation of the Burgos government.[8]

Despite having officially accepted the non-intervention, in practice Portugal spared no efforts to support the nationalist cause. And although there was no formal military intervention, the support it granted was not at all irrelevant for the outcome of the conflict.[9]

Diplomatic support, to begin with. Both at the Non-Intervention Committee and at the League of Nations (LN), it was the Portugal of the *Estado Novo* who relayed, internationally, the positions of nationalist Spain. So much so that a number of historians have considered Salazar to have been, not only the *Estado Novo*'s Minister of Foreign Affairs, but also General Franco's minister for foreign policy.[10]

Secondly, logistic support. Ports, airports, roads and railways represented pivotal points in the supply traffic of weapons, ammunition and other military equipment originating in the European powers and whose immediate and unchecked recipients were the nationalist forces. On the other hand, with the knowledge and consent of the *Estado Novo*, it was within Portuguese territory that armament deals were contracted between Franco's agents and international suppliers. Not to mention the Portuguese companies that accepted the role of intermediaries between the nationalists and European and American companies, and the financial conveniences in credit granting by Portuguese banks to the Burgos government.

Thirdly, political support. Portugal adamantly shut its door to republican forces, thus preventing that the Portuguese territory might be used as support base or even shelter to the "reds", whereas the nationalists benefitted from the tolerance of Portuguese authorities and, more than that, of the support from the media, namely the Portuguese radio broadcaster, which constituted an important means of political propaganda in the service of the nationalist cause.[11]

Finally, military support, which did not translate into the formation of any Portuguese expeditionary corps, but instead took two different shapes: the dispatch of a Portuguese Military Observation

Mission in Spain and the recruitment of volunteers to fight alongside Franco's troops: the "Viriatos".

The Portuguese Military Observation Mission in Spain is put together in March of 1937, with the encouragement of General Raul Esteves, and it had a three-fold purpose. The first was essentially technical-military: observation and contact with new technologies and warfare techniques and procedures; the second was political-military: to provide support and assistance to the Portuguese volunteers fighting in the nationalist ranks; the third was primarily political: to secure a political position for Portugal in the peninsular and perhaps also the European context.

Neither an expeditionary corps nor even an exclusively Portuguese force was ever created, and so the majority of the Portuguese military was eventually incorporated in the different units of the Foreign Legion or the Francoist army. Estimates indicate that, through the Portuguese Military Observation Mission in Spain, whose head of support section was, since March of 1938, Captain Jorge Botelho Moniz, passed a total of 164 military from the Portuguese Armed Forces (153 from the Army and 11 from the Air Force).[12]

The expression "Viriatos" was popularised by the press and the propaganda favourable to General Franco and designated the Portuguese soldiers that fought under the flags of the Foreign Legion, in the Falange and the Requetés militias or in the various regular units of the Army, during the Spanish Civil War. They signed up individually and voluntarily, and were afterwards integrated and deployed to the battlefront. As always happens in wartime, numbers are themselves the object of propaganda, but according to recent estimates, around 2,654 "Viriatos" are supposed to have fought on the side of nationalist Spain".[13]

It might be surprising to know, but there were also Portuguese fighters on the republican side. They did not leave for Spain after unofficial recruitment and integration by Portuguese authorities, but rather signed up individually or through networks of political solidarity and were for the most part either Portuguese economic emigrants, whose numbers had grown since the 1920s, or political exiles with different ideological orientations – republicans such as the Buddhas, socialists, communists and anarchists – escaping the repression of the Portuguese dictatorship. They swelled the ranks of socialist and anarchy militias and even those of the Republican Army. Here also numbers are controversial, but according to the same estimates, they have risen above 1,000.[14] After the war, defeat escorted them to the same fate as their comrades: flight, imprisonment or execution by firing squad.

The Portuguese strategy of ambiguity and diplomatic duplicity was successfully handled during almost the whole course of the civil war. Only on one occasion, during the Munich crisis of 1938, did the convergence between Portugal's official and non-official position run any substantial risk.

The *Estado Novo* had broken off diplomatic relations with Spain in October of 1936. In July of 1937, the Portuguese Consul General settles in Salamanca and, in December, Salazar appoints a "special liaison" (the future Ambassador Teotónio Pereira) of the Portuguese government to Franco's, which meant, politically and diplomatically, the recognition *de facto* of the Burgos regime. Formally, however, he will wait for London's *agréement* and will only give his recognition *de jure* after England itself, in April of 1938.

In this stage, the fate of the war was practically decided and the concerns of the *Estado Novo* shifted to a new threat: the danger of Iberianist annexationism by a number of Falange's more radical fringes and the possibility of a future alignment between Franco and the central powers, hostile to England. Salazar's great concern was now stabilising diplomatic relations with Spain and ensuring Portugal's neutrality in the foreseeable European conflict. A condition he knew was crucial if he intended to persist in staving off the choice between the two imperatives and to ensure Portugal's geopolitical balance.

To this purpose, towards the end of the war in Spain, Salazar refrained as much as he could from taking political advantage of Franco's victory, and reduced Portuguese presence at the ensuing celebration party in the streets of Madrid to a minimum. Moreover, he clearly and overtly lost interest in the fate of the "Viriatos" after the demobilisation. It was the last exercise in his strategy of ambiguity, a strategy which he in fact manoeuvred successfully.

The end of the civil war and Franco's victory created the perfect conditions for the materialisation of Salazar's notion of a "peninsular friendship", the beginning of a period marked by a good relationship and political solidarity between the two authoritarian regimes of the Iberian Peninsula.

The good relationship and the political solidarity had their diplomatic substantiation in March of 1939, with the signature of the Portuguese–Spanish Treaty of Friendship and Non-Aggression – the so-called Peninsular Pact or Iberian Pact. With a view to reinforcing its terms, an additional protocol was signed between the two countries in July of 1940, after the Fall of France, in the early months of the war.[15] With the signature of the Iberian Pact and the Anglo-Portuguese Alliance, which it sought to reinforce, Salazar held in his hands the two chief diplomatic instruments of the country's geopolitical balance, instruments that would lead Portuguese foreign policy throughout the Second World War.

By the end of the Spanish Civil war, Salazar had managed to accomplish his two great political objectives: to safeguard the State's external security and the internal consolidation of the regime.

The Spanish Civil war had weighed heavily on Portugal's foreign policy, but no less on internal radicalisation and the regime's fascistic drift. Both entailing profound consequences for defence policy and the reforms of the military instrument.

From internal threat to continental threat: Defence policy and military reforms

The military coup of the 28th of May of 1926 had overthrown the Republic and installed a military dictatorship. It did not, however, end the financial crisis or the political instability. As it might be expected, the first priority of the new regime was its political affirmation and consolidation.

In the second half of the twenties, a concrete and clear-cut external threat was not yet on the horizon. The international strategic atmosphere was one of peace and security, the "spirit of Geneva" ruled and the League of Nations (LN) was living its golden age. For the military dictatorship, on the contrary, the concrete and clear cut threat came from within. And although it never amounted to more than a veiled menace, its shadow pervades all of the defence policy of the military dictatorship. The political consolidation of the regime was top priority and, in this sense, the first military reforms undertaken by the Dictatorship had two purposes: to endow the country with a military instrument that was at once technically effective and politically loyal to the regime. Military effectiveness and political control are then merged in these military reforms that are comprised of four key areas: first, the model of military recruitment; second, organic and territorial structure; third, reduction of military personnel; and fourth, quipment and armament.

As such, in the second half of the twenties, between 1926 and 1930, the Army was top priority and the primary goal of the Dictatorship was the dismantlement of the military system inherited from the First Republic. Immediately after the coup of the 28th of May, in June of 1926, abundant legislation was passed on the general reorganisation of the Army. The new law refused the Swiss-like "militia system" adopted by the Republic and instituted a "mixed system" which, while preserving conscription, reduced the militia contingent and reinforced the professional cadre. The general principle of the universal and compulsory military service still held, but was not enforced. In practice, the legal possibility of paid exemption affected radically the universality of conscription, favouring higher social strata. At bottom, it meant a return to the old system of "remissions" in effect during the Monarchy, only now without the former designation. Concerns over personnel and expense cuts also led to a reduction in the length of military service: 12 months according to the legislation of 1926, increasing to 17 months in 1927 and, as a compromise solution, finally fixed at 15 months from 1929.[16]

The military policy of the dictatorship also discarded the republican legacy at the level of organic structure and territorial organisation. Territorially, the mainland was divided into four military regions (Porto, Coimbra, Tomar and Évora), added by the Military Government of Lisbon. The Azores and Madeira, formerly military commands, also become military governments. The arms' structure was also the object of significant reforms. The first of these, by virtue of the growing strategic importance of the Air Force during the Great War, was the creation of the Military Aeronautics as an autonomous service within the Army. But the traditional arms, now endowed with autonomous leadership, also underwent new reforms: in infantry, the creation of the battalions of *caçadores* ("hunters"), machine guns and mortar batteries; in artillery, the creation of heavy and campaign artillery units, as well as several depots; and in cavalry, the creation of cavalry brigades.

These changes in recruitment and organic structure did not imply an increase in the number of personnel. Quite to the contrary, in a scenario of crisis in public finances, the objective was resizing the military instrument, still oversized following the Portuguese military effort in the First World War.[17]

As to the major concern of the Armed Forces – the much needed rearmament – once more the crisis in public finances precluded the implementation of any modernisation policy worthy of that name.[18]

Apart from a number of contingency measures, the Navy did not undergo any in-depth reforms during this first period of the military dictatorship.

The affirmation of a defence policy at variance with the Republic was not directly merely at the strictly professional domain of the technical-military sphere, but also, and primarily, to the political control of the Armed Forces. To this purpose, the military units whom the government did not fully trust politically were moved away from the capital or simply extinguished, namely, following the succession of botched coups against the Dictatorship. On the other hand, elite units were created whose loyalty was with the regime, as well as battalions of *caçadores*, machine gun battalions and cavalry brigades, strategically allocated across the national territory and in the accesses to the capital. This policy is supplemented by a series of social-professional measures intended to garner the military's political support for the regime: above all, a policy of promotions and pay rises for officers.[19]

Guided by the same goal – political control – internal security forces, which had been created by the Republic and were still ruled by republicans, were also refashioned. The Republican National Guard, casted off to the rural areas and scattered across the countryside, the Criminal Investigation Police, reorganised from top to bottom, and the State Security Police was simply extinguished.

From 1930, in the absence of any discernible change in threat perception and in the orientation of military policy, there were, however, a number of factors that shifted the priority from the Army to Navy. At the level of internal policy, the regime was undergoing consolidation and a transition process was commencing from a military dictatorship to a civil dictatorship. On the other hand, in the Atlantic islands and in the colonies, rumours circulated and the threat of uprisings against the Dictatorship was in the air. One such threat actually materialised – the Madeira uprising of 1931. In this context, mobility became a crucial strategic factor and the Navy a military instrument of the utmost importance. On the other hand, the Armada was historically the branch of the Armed Forces most inclined to republicanism and the less politically trustworthy. In a sense, the priority awarded to the Navy was part of a policy designed to win its support, or at least to neutralise its hostility. For all the stated political reasons, but also for technical and financial ones – since no rearmament plan for the Army seemed to gather a consensus and, should one exist, such a plan would prove to be considerably more expensive – in 1930 an ambitious plan for naval rearmament is launched. A plan based on Pereira da Silva's project and conducted by Minister Magalhães Correia. Between 1930 and 1935, the Navy acquired 14 new ships, among which three destroyers and his first three submarines. However, the plan would not advance beyond its first phase, being suspended in 1935.[20] Despite the interruption, the plan accomplished, in the short run, the government's two-fold objective: on the one hand, the Armada managed to substantially increase its tonnage (from 19,168 t in 1928 to 31,254 t in 1936, that is, 63%) and procured a minimal yet effective capacity for navigation protection and coastal defence; on the other, it facilitated the neutrality, if not the support, of the Navy officers, demonstrated by the so-called "sailors' revolt" in 1936. A revolt in which not a single officer took part and which was circumscribed to a few units and promptly crushed.

By the end of the Republic, military expenditure (Army and Navy as a whole) amounted to 21.2% of the total public expenditure. This number rose slowly and gradually throughout the period, standing at 22.8% in 1930; peaking in 1933 at 27.6%; and registering 25.4% in 1935. The relative weight of the two branches evolved also in consonance with the fluctuation of priorities. Until 1930, the Army represented almost 70% of the expenditure, whereas the Navy neared the 30% (68.8% against 31.2% in 1930). From 1930, with the Navy's modernisation plan, the branches present an almost balanced share of the expenditure (the Army's 53.2% against the 46.8% of the Navy). The trend is once again reversed, gradually, until the plan is suspended in 1935.

Now, 1935 marked precisely a turning point in priorities: the end of naval and the return to terrestrial primacy. The reason was simple enough: the end of naval priority had been dictated by the insurgencies, in the islands and the colonies, against the Dictatorship. The *Estado Novo* was consolidated and a new kind of threat had emerged and loomed over the country. It was no longer an internal menace emanating from the overseas possessions, but an external threat coming from the mainland. It was the return of the "Spanish peril", in which were merged the State's external security and the internal survival of the regime.

With the implantation of the Republic in Spain, especially in the "red biennium" of 1931 to 1933, and after, with the triumph of the Popular Front in 1936, the continental threat took a definite shape for Portugal. Whether it was an offensive from Spain, border incidents resulting from the war, the most likely scenario would always be that of a continental threat. The Spanish Civil War corroborated that trend and marked the beginning of a new phase for defence policy and for the reorganisation of the military instrument in Portugal. A new defence policy begins to take form, one which the *Estado Novo* will implement in stages with a constant strategic goal in mind: to fight the continental threat.

In 1936, Salazar took over the Ministry of War, which is in itself tremendously revelatory of the internal and external importance, for the regime, of the subject of defence. Salazar outlined four strategic guidelines: stable economy and public finances as a basic condition for total war; modernisation of the metropolitan army to ensure the defence of the mainland; the preservation of the colonial army; and naval forces able to ensure Communications and the reinforcement of defence in the Atlantic islands, the colonies and strategic ports. To carry out his defence policy, Salazar summoned a young captain who will later become Under-Secretary of State of War, Minister of War and Minister of Defence: Fernando Santos Costa. He will be a pivotal element not only in military reforms but also in the relations between the military and Salazar's regime.[21]

Although Salazar's defence policy encompassed four vectors, the shadow of the continental threat, aggravated by the Spanish Civil War, shifted priorities from sea to land and, consequently, from the navy to the Army. Ever since 1935/1936, an ambitious modernisation plan had been approved and set in motion which, according to three threat levels, should be directed towards three different objectives. The first threat level included actions of the opposition starting in the border; the second, the onset of the civil war and the ensuing likelihood of hostile actions in the Portuguese border; and finally, the third and most serious, the chance of a Spanish invasion.[22] Hence, the plan should be carried out according to three stages: the first had as its goal the immediate defence of the border by a modern division; the goal of the second was to ensure the defence of Lisbon during the time required for the arrival of foreign military assistance – that is, *bien entendu*, provided within the scope of the Anglo-Portuguese Alliance, which should be composed of a four-division force; and finally, the third, whose objective was the containment of an attack from Spain, with a minimum or no outside aid at all, which would require the creation of a military force of gigantic proportions, given the Portuguese strategic potential. No more and no less than ten divisions.

This plan is complemented, in 1937, by a new reform headed by Santos Costa and enshrined in the new army reorganisation law of 1937, which introduces alterations at three levels: recruitment system and military service; organic structure; and civil-military relations, that is, between the Armed Forces and politics.

According to the strategic concept that prevailed in the "nation at war", which since 1914–1918 had replaced the Napoleonic concept of the "nation in arms", the recruitment system and military service remained compulsory. Its underlying idea was that of a mass army, albeit preserving the principle of legal exemption, that is, the old system of remissions. Length of service increased from four to six years. Military training had a duration of four months, followed by one year of effective service, by the end of which the soldier integrated the reserve availability until the six years were completed.[23]

As to organisation, the Army was now divided into two distinct, albeit ruled by common organic principles, forces: the metropolitan army, wherein only metropolitan Portuguese citizens were allowed to serve, and the colonial army, wherein not only Portuguese citizens from the metropole could serve, but also "the natives" from their respective colonies. As regards personnel, there is also a clear intention to implement cuts, motivated by two chief concerns: decrease in expenditure and a greater concentration of resources in order to maximise operational effectiveness. In infantry, which suffered the greatest cuts, the total number of regiments declined from 22 to 16. On the other hand, during peacetime, troops were now divided between border units and line units. And, except for minor changes in the areas of responsibility, the territorial division of the mainland was preserved. One other novelty was the creation, under direct dependence of the minister of Defence, of a Command General of the Aeronautics, which mirrored the growing importance of the Air Force.[24]

In the same spirit of operational modernisation and with the purpose of raising the standards of the Portuguese Army so as to bring them closer to those of the other European armies, especially since the arms race by Nazi Germany, new units are created: one tank battalion, three motorised rifle battalions, one motorised cavalry regiment with automatic machine guns squads, motorcycle squadrons and motorised squadrons.

However, the great organic change undertaken by the 1937 reform was the creation, within the Army, of a separate Staff corps. The military dictatorship had already implemented a policy of pay rises and quicker promotions for the Army Staff, and now the *Estado Novo* reinforced that policy, turning it into a separate corps within the officer corps. This separation, whose technical purpose was the division of functions between strategic conception and operational execution, answered also to a political agenda and had political consequences.[25]

Finally, the relationship between the Armed Forces and politics and internal security. Ever since 1936, in the context of the Spanish Civil War, there had been a radicalisation of the regime and a fascistic drift that led to a reform of the repressive apparatus with the establishment of a political police (PVDE), but also the creation of new paramilitary institutions with political mobilisation and civil defence duties, such as *Mocidade Portuguesa* ("Portuguese Youth") and the *Legião Portuguesa* ("Portuguese Legion"). Although autonomous from the Ministry of War, there existed, between these paramilitary institutions and the Armed Forces, a number of informal mechanisms of dependence that thereby connected the military institution to the political consolidation of the regime.[26]

Santos Costa's rearmament programme itself, from 1935–1936, by satisfying the officer corps' traditional claims for Army modernisation, also alienated the officers from political intervention, thus contributing, in an indirect manner, to the consolidation of the regime. However, the Spanish Civil War and the vicissitudes of the diplomatic relationship between Portugal and England hampered and delayed British military aid,[27] forcing the government, from 1938, to progressively turn to German armament, indeed with no great success.[28]

The Military Aeronautics would have to wait for the fifties to affirm itself as an autonomous branch of the Armed Forces. So far, it was not considered a priority, and the prevailing strategic conception was still based on the traditional view of the Air Force as an auxiliary weapon either of the Army or the Navy. In the first half of the decade, modernisation efforts were directed primarily to the naval branch, in the scope of Pereira da Silva's plan; in the second half, to the Army, in the scope of Santos Costa's reform.[29]

Between 1936 and 1939, the budget for defence increased slowly and gradually. In 1936, military expenditure amounted to 26.7% of the total public expenditure. In 1939, in the beginning of the war, it stood at 33%. According to the primacy awarded to the Army, in the stage of the defence policy, the navy declines in the share of expenditure. If, in 1934, towards the end of the plan for naval rearmament, Army and Navy were almost even and the Armada represented 44% of expenditure, as early as 1936 it had fallen to 30% and continued to decline progressively until de Second War.

In short, if Salazar's and Santos Costa's reform had been successful in its strategy to politically control the military instrument,[30] the same could not be said of its outcome as regards modernisation and operability.

In 1939, by the end of the Spanish Civil war, the state of the Armed Forces was less than splendid. The 1930 naval plan had been suspended in 1935 and, along with it, the modernisation of the Navy. And of the ambitious Army reform plan of 1935–1936, one single modern division was up, there were grave insufficiencies in heavy armament and, at best, only three days' worth of ammunition.

The condition of the Armed Forces would soon represent a weighty constraint for the Portuguese position in the fast approaching Second World War.

CHAPTER

5

The *Estado Novo* and the Second World War

Portugal's international standing and foreign policy on the eve of the Second World War

The Spanish Civil War was still unfolding when, from the summer of 1938, the crisis became more acute and the signs of war in central Europe amassed. Nazi Germany persisted in successive shows of force, which neither the Munich Conference nor the "appeasement" policies of Western democracies managed to curb. The invasion of Poland, in September of 1939, triggered the breakout of a war that had been impending since the previous year.

In this new strategic environment, at the check-out of the Spanish Civil War and on the doorway of the Second World War, what was Portugal's position internationally? What were its national objectives, its diplomatic instruments and in what condition were its Armed Forces to face the conflict that had just started?

Portugal's international position did not differ much from that which, historically, it had occupied in other international conflicts, constrained as it was by the same geopolitical constants: in Europe, the preservation of peninsular balance and the strategic importance of the Atlantic archipelagos; overseas, the integrity of the African empire.

However, the international standing of the *Estado Novo* on the eve of the Second World War was, in spite of all, less serious than that of the Republic at the threshold of the First World War. Primarily because the international conjuncture was such that a small power was allowed to enjoy, within its own limits, a greater autonomy in its foreign policy. The reasons for this were simple and practically the opposite of what had occurred in the First War. In the first place, there was no "spectre of Cartagena" looming in the peninsular context, as there was from 1907 and until the onset of the First War. Meaning that, not only there was no diplomatic rapprochement between London and Madrid, but, moreover, that the two countries stood, internationally, on opposite ideological sides of the European conflict. Secondly, there was no naval Anglo-German rivalry in the Atlantic, as had been the case before the First War, and not only England ruled the Portuguese Atlantic space, it was also free to do it without the need to rely on the Portuguese archipelagos, since it had alternative bases in Gibraltar and the Bermuda. Finally, in third place, Germany had lost its colonial empire and there was no standing deal – secret or not – concerning a partition of the Portuguese colonies, as in 1898 or 1912–1913.[1] It made all the difference for Portuguese foreign policy.

Also, domestically, the circumstances of the *Estado Novo* were far more comfortable than those of the First Republic. Whereas the Republic had never been able to overcome the issues surrounding governmental instability, political consolidation and the regime's national legitimisation, or to attain nationwide consensus on its position vis-à-vis the war, not only had the *Estado Novo*, on the contrary, secured political consolidation, the Spanish Civil war had furthermore allowed for a strengthening of the regime. The dictatorship, in its turn, favoured – not to say imposed – a strained consensus. Additionally, whereas the Republic had to face a hostile monarchy in Spain, the *Estado Novo* profited from a favourable and politically like-minded regime, whose establishment, moreover, it had supported during the civil war, and with whom it had signed a pact by the end of the war.

Does it mean that circumstances were untroublesome? Indeed not. There were many constraints and little leeway, in between Spain – on which the regime relied – and the Anglo-Portuguese Alliance, on which the Empire depended.

In the likelihood of a conflict, the national goals were, *mutatis mutandis* the international conjuncture, the historical goals of Portuguese foreign policy: in Europe, the external security of the State, that is, national Independence, which entailed the preservation of peninsular duality; overseas, the integrity of the Empire. To these external goals was added, at the domestic level, the survival of the political regime.

Now if, during the First War, these goals had led to belligerence, in the Second World War they will impose neutrality.

In pursuing its objectives in the course of the Second World War, Portuguese foreign policy, with Spain on one side and England on the other, will make use of the two diplomatic instruments it had created and which were now at its disposal: the Anglo-Portuguese Alliance and the Iberian Pact.

The former was Portugal's historical alliance. An alliance in which, after a short period of rather strained relations between Portugal and

England, during the Munich crisis, Salazar endeavoured to reinvest and reinforce, refocusing its foreign policy and reconnecting with London from the mid-1938.

The latter represents the corollary of the relationship between Salazar and Franco during the Civil War. Towards the end of the conflict, when the nationalist victory was all but decided and in the face of an impending European crisis in which the two states might become involved in opposing belligerent sides, the Burgos government proposed the signing of a friendship and non-aggression treaty between them. Salazar was aware of the importance of the Anglo-Portuguese Alliance and of its constraints. Cautiously, he did not accept a peninsular pact that might override the Anglo-Portuguese Alliance in the priorities of Portuguese foreign policy. He imparted his conditions to Franco, deferred the decision and consulted with England. In case of a European conflagration, a neutral Iberian Peninsula suited England well, and so it gave Portugal its approval. For Portugal, the strategic neutralisation of the peninsula and the reciprocal neutrality of the two peninsular states were crucial and tantamount to a two-fold assurance: of the "peninsular friendship" and, consequently, of Spain's commitment to respecting common borders, that is, Portugal's national independence. Formally the Treaty of Friendship and Non-Aggression, and later known as the Iberian or Peninsular Pact, was signed in March of 1937.[2]

Thus, Portugal's position with regard to the onset of the Second World War, in September of 1939, was clear and had been well defined. Its great strategic goal was neutrality and the two diplomatic instruments to secure it were the Anglo-Portuguese Alliance and the Iberian Pact.

Why neutrality: Portugal and the Second World War

The invasion of Poland, on the 1st of September of 1939, and the successive declarations of war that ensued, marked the onset of a world war that would last for six long years until the 2nd of September of 1945. Immediately following the invasion of Poland, unhesitatingly and even before England had declared war on Germany, Salazar solemnly declared Portuguese neutrality. Three major reasons – one of them historical and the other of a political, diplomatic and military nature – motivated that position: The historical memory of the Great War which he did not wish to see repeated, the Republic's ambiguous and undeclared neutrality, between 1914 and 1916, and the active belligerence of 1917 to 1918. More than that, the neutral stance was prompted by the national goals mentioned above, and the imperatives of the moment. Firstly, the immediate diplomatic need to draw away Francoist Spain from the Axis powers and urge it to commit to neutrality as well. Secondly, the military constraint represented by the state of the Portuguese Armed Forces, which Salazar knew to be unprepared to fight a war.

Yet, what kind of neutrality and why neutrality in the first place? From September of 1939 to April of 1945, that is, for almost all the duration of the conflict, Portugal maintained its formal position of neutrality. The political nature of its neutrality, however, did not remain unchanged. On the contrary, it evolved with the evolution itself of the interplay of interests of the great powers and the war's military strategies.[3]

Throughout the first phase of the war, marked by Axis supremacy, Portuguese neutrality experienced two distinct moments.

The first moment lasted between the start of the war and the summer of 1940. During this period, the Iberian Peninsula and the Atlantic space did not appear to be strategically important for the conduct of war, and none of the great powers exerted diplomatic pressure for the Iberian states to declare belligerence. Neither did Germany requested it from Spain or England, to whom the strategic neutralisation of the peninsula had always appeared as convenient, from Portugal. In that context, both peninsular states, Portugal and Spain, declared neutrality and were further convinced they could enjoy a degree of autonomy vis-à-vis the two nearest belligerent sides. In this first moment, Portugal's was a strict neutrality, although benevolent to England.

The second moment lasted from the summer of 1940 to the summer of 1941, when the Axis supremacy was still in place. There was never a period when the evolution of war strategies and the clashing goals of the belligerents represented a greater threat to Portuguese neutrality.

The German troops, already prevalent in central Europe, essayed, as an exercise in naval strategy, an offensive against southern Europe with the purpose of taking control of the Mediterranean, northern Africa and the Atlantic islands. To prepare against such an eventuality, which would strategically entail the arrival of the Germans in Gibraltar or their stationing in the Canary or the Cape Verde islands, jeopardising British communications with the south Atlantic, England developed plans for a preventive establishment of premises in the Azores.

Now, in this scenario, Franco's Spain, which had declared its neutrality and maintained it during all of the *drole de guerre*, fearing a French invasion against which it felt powerless, rapidly evolved to a position of near alignment with Germany. After the fall of France and in view of Germany's increasingly likely victory, Franco, concerned with the possibility of being excluded from the table of the victors, took a stand regarding Portugal. He declined Salazar's proposal of a joint declaration of neutrality and proclaimed his "non-belligerence". He was not belligerent, but also not neutral. It had the outward appearance of a diplomatic subtlety but possessed a clear political meaning. The day was the 12th of June of 1940 and Spain had one foot in neutrality and the other in belligerence. It stood one step away from entering the war. On the side of the Axis, that is.

Spanish belligerence now suited Germany, as being the key to reach Gibraltar, the Mediterranean and North Africa. But only if there were no previous conditions attached to it. To Spain, on the other hand, belligerence was convenient, as long as a number of conditions – stated from the outset – were fulfilled. Territorial conditions, to begin with: Gibraltar and a North African empire. A period of Spanish-German negotiations hence began which dragged on diplomatically with no satisfying end in sight.

For Portugal, it all had tremendous consequences. Spain's entry into war beside the Axis powers would force Portugal into entering on the side of the Allies. In such a scenario, Portuguese foreign policy would be left with no other choice. In an effort to prevent it, Salazar reinforced Portuguese neutrality in stricter and more equidistant terms, as a way to pressure Franco to declare Spain's neutrality.

Spain played in the opposite direction. It attempted to lure Portugal into the sphere of influence of the Axis, which was also its own, and pressured Lisbon to withdraw from the Anglo-Portuguese Alliance with the prospect of an Iberian military pact against England.

For Portugal, this was an unacceptable proposal. It overtly challenged Portugal's very objectives, which essentially relied on the Anglo-Portuguese Alliance: both the peninsular balance and duality and the integrity of the colonial empire. As befitted his position, Salazar refused and countered by proposing an alternative solution: the reaffirmation and broadening of the Friendship and Non-Aggression Treaty based on a principle of collaboration that ensured respect for and the inviolability of their "respective metropolitan territories".

Salazar's counter-proposal was accepted by Franco and was soon after substantiated in an Additional Protocol to the Treaty of Friendship and Non-Aggression, signed in Lisbon in July of 1940.[4]

Hence Portugal achieved a significant diplomatic victory and, with it, a double assurance: on the one hand, the reinforcement of the Anglo-Portuguese Alliance and, furthermore, England's avowal that it would not resort to Portuguese territory, namely the Azores. On the other hand, Spain's assurance with respect to the security of the Portuguese border, which, in its turn, made progressively more difficult and remote the possibility of Spanish belligerence.

Throughout the war, Portuguese foreign policy was never as strictly neutral and equidistant as during this period. It was the period of the so-called "geometric neutrality", marked by the greater importance awarded to the peninsular alliance. Portugal's neutrality, in those terms, eventually influenced Spain's withdrawal from belligerence. The strategic evolution of the war accomplished the rest.

Diplomatic negotiations between Germany and Spain dragged on until October of 1940. On this date, Hitler and Franco met in Hendaye, but, contrary to expectations, the deadlock was not dissolved.

Towards the late-1940, there were plans for a German offensive in southern Europe – the notorious Operation Felix – with the intent to control the Mediterranean, take Gibraltar and deny the Allies access to the Iberian Peninsula and the Atlantic islands. It had been scheduled for January of 1941, but in December of 1940, all of a sudden, the strategic priorities of Nazi Germany changed. Hitler decided that east, rather than south, was top priority. He cancelled Operation Felix and moved forward with Operation Barbarossa.

Germany's strategic interest in the Iberian Peninsula dwindled, as the East took precedence and the Soviet Union was invaded. Germany had now become one more interested party in the neutralisation of the peninsula. Portugal's neutrality was no longer a concern.

It is unarguable that, during the entire period, Salazar's policy of "geometric neutrality" and the importance awarded to the Iberian alliance were important factors for neutrality in the Iberian Peninsula. However, they were neither the only, the decisive, nor even the most important factors involved. What enabled and ultimately favoured neutrality was the strategic evolution of war itself. That is, England's interest in the neutralisation of the peninsula, Germany's inability to develop a naval strategy against England, which would entail the use of the Atlantic archipelagos and, finally, the shift in Germany's strategic priorities. If Germany had strategically opted for Operation Felix, that would have implied not only the invasion of the Portuguese territory, but also England's occupation of the Atlantic islands. Having opted for the Operation Barbarossa and the invasion of the Soviet Union, Germany's interest eventually coincided with England's in respect to the strategic neutralisation of the Iberian Peninsula and the neutrality of the peninsular states.[5]

During the second phase of the war, marked by a balance of

strengths between the belligerents and lasting between the summer of 1941 and the spring of 1943, Salazar kept unchanged Portugal's stance of "geometric neutrality". However, as the scales began to tip on the side of the Allies, especially after the entry of the United States into war, also in the scales of Portuguese foreign policy, the peninsular alliance loses weight in the inverse proportion of the Anglo-Portuguese Alliance's gain.

During this phase, while Germany proceeded with its invasion of the Soviet Union, the Allies began to plan an attack against French North Africa – the Operation Torch. Now, the neutrality of the Iberian Peninsula was key to the success of this operation, seeing that a belligerent Spain might constitute a strategic obstacle.

In December of 1941, the United States entered the war and the conflict acquired a global dimension. From that moment, the war scales began to tip on the side of the "United Nations". This is the context wherein Francoist Spain commences to draft a diplomatic rapprochement towards Portugal. It is Franco who requests a meeting with Salazar. It takes place in Seville, in February of 1942, and marked the end of two long years of bilateral strain between the two Iberian countries. Spain abandons non-belligerence and draws near Portugal's "geometric neutrality". The rapprochement is reiterated by the visit to Portugal of the new Foreign Affairs Minister, who is favourable to neutrality – Count Jordana – and the proclamation of the peninsular or Iberian Bloc in December of 1942.

In November the Allies had launched Operation Torch and the occupation of the French territories in North Africa. It amounted to a radical change in regional strategic balance.

Throughout this period of the war, Portugal's "geometric neutrality" continued to influence Spain, which eventually adopted the same position. Important as they were, the stances of the Iberian countries were not as decisive for peninsular neutrality as the global evolution of the military strategy of the war: the precedence of the Soviet Union, for Germany, and that of North Africa, for the Allies. The neutralisation of the peninsula suited both belligerents.[6]

In the final phase of the war, marked by the primacy of the "United Nations", the scales were already tipping on the side of the Allies. Portuguese foreign policy accompanied the evolution of the war's shifting strengths and went through two distinct moments until the end of the war. First, it evolved from "geometric neutrality" to "cooperative neutrality", which also meant that the primacy of the peninsular alliance gave way to that of the Anglo-Portuguese Alliance. In a second moment, matching the end of the war, Salazar, who had always been reluctant towards the opposite shore of the Atlantic, relaxed his position and Portugal evolved towards cooperation with the United States, which would be further consolidated in the aftermath of the war.

The first moment unfolded between the autumn of 1942 and the spring of 1943 and lasted until the summer of 1944. During this period, the strategic situation of the Iberian Peninsula changed significantly due to two reasons. On the one hand, the allied victory and the expulsion of the Germans from North Africa, in June of 1943, which extended the neutralisation of the Mediterranean coast to Vichy's France.[7] On the other, the Allies' strategic decision to open new fronts in Sicily and Normandy. Both converted the Iberian Peninsula into an extraneous and strategically irrelevant theatre of war, dispelling once and for all the possibility of a German invasion.

This evolution had a profound meaning for Portugal. It meant that, to Spanish neutrality, was now added the strategic neutralisation of the Iberian Peninsula itself – which had lost any military purpose in the conduct of war. That was assurance enough of the definite establishment of peninsular neutrality and enabled Portuguese foreign policy to evolve in its positions and to diversify its objectives.

Hence, diplomatically, it evolved from "geometric" and equidistant neutrality to "cooperative neutrality" – with the Allies, that is. This evolution was naturally accompanied by a shift from the circumstantial primacy of the peninsular alliance to a progressive return to the structural primacy of the Anglo-Portuguese Alliance.

At the level of objectives, evolution was no less significant. Having assured the chief objective of duality and peninsular stability, the *Estado Novo* progressed towards new and increasingly more important objectives: economic interests, the colonial integrity of the empire, and, anticipating the near future, the survival of the regime in the post-war period. The fate of the war all but decided, it was all very clear: only the Western Allies could guarantee their fulfilment.

In the global strategic scenario, the victory in North Africa and the neutralisation of the Iberian Peninsula altered the strategic priorities, which were shifted to the Atlantic. In such a context, England and the United States finally became interested in making use of the military bases in the Azores. For Portugal, that decision would prove vital.

The Anglo-American strategic priority coincided, in time, with the return and reinforcement of the Atlantic vector of Portuguese foreign policy. In June of 1943, England formally requested of Portugal the concession of military facilities in the Azores archipelago. The Portuguese response was swift and positive. In October of 1943, the agreement was signed and within the same month the British troops arrived in the Azores bases. Spain, having been consulted – under the

Iberian Pact and the Additional Protocol – the day before, did not oppose to its terms and even Germany only went so far as to make an inconsequential diplomatic remonstration.

It was the beginning of Portugal's "cooperative neutrality" and, with it, the confidence in British assurance regarding the endurance of the Portuguese colonial army. But the war was not over, and the assurance was not solid, let alone definitive. And there was still a more difficult and uncertain objective left to fulfil: the political survival of the regime. The survival of a dictatorship in a world in which foreseeably democracies would prevail. Finally, there was still the question of defining the relationship with the United States, whose predominant role in the outcome of the war warranted its international protagonism in peace management.

The second moment began in the summer of 1944 with the invasion of Normandy, and lasted until the end of the war. During this period, Portugal shown signs of reinforcing its "cooperative neutrality" and launched, under diplomatic pressure and with active support from the Anglo-Portuguese Alliance, a rapprochement to the United States of America. Not without resistance from Salazar. His main concern, though, was prudent and defensible: to safeguard national objectives in post-war management.

In the last phase of the war, the Atlantic had acquired strategic prominence and the United States had emerged as a great naval power. Control over the Atlantic had become key to its strategic interests. Hence the diplomatic rapprochement was Washington's initiative, and aimed at ensuring, beside and outside England's influence, exclusive and permanent use of a military base in the middle of the Atlantic: in the Azores. The Lajes field, occupied by England since October of 1943, stipulated aircraft stopovers and refuelling, but did not allow permanent stationing of North American military units. Washington requested, in the first place, the Santa Maria base, to which it intended to award a new strategic role. It was no longer the traditional naval role of providing support against submarine war, which, in the final phase of the war, the Allies had already won, but a new role in air support. In the middle of the ocean, the Azorean base would become a sort of natural aircraft carrier, the central point of logistic support for the gigantic air bridge assembled by the United States in the direction of Europe, which later on and in a different context, would be enshrined in NATO's doctrine as the "rapid reinforcement of Europe".

Salazar was never sympathetic to the *American way of life*, which was diametrically opposed to his world view. And he always looked upon North American presence in Europe with distrust, fearful as he was that it might foster the democratic principle of "the peoples' right to dispose of themselves", that is, democratisation and decolonisation. Hence Portuguese diplomacy was initially inclined to refuse the American request and deferred its response, on the expectation of England's support. However, relations between Portugal and its ally of old were crossing a period of greater strain caused by the Portuguese reluctance to declare an embargo on wolfram sales to the central powers. During the "wolfram question", England went so far as to support the military conspiracy against the *Estado Novo*. But from the moment when Portugal gave in on this point, in June of 1944, things went back to normal and England dropped the opposition and ensured the continuity of Salazar's regime.

Without formally relinquishing neutrality, Portugal intensified its collaboration with the Allies and essayed a response to the United States. Not directly to Washington, but via London. It was still hopeful that the British might back its claims, even if unofficially. Their reply, however, was quick, univocal and negative. England considered the American presence in Europe to be fundamental, not only for achieving victory in the war, but also for the containment of the Soviet Union in its aftermath. Portugal is left with no alternative but to negotiate with Washington. These negotiations resulted in Salazar's granting the United States, albeit only provisionally, permission to use the Santa Maria base.

In that manner, Portugal secured a number of trade-offs of great relevance for its new colonial and political objectives: in the first place, colonially, the assurance of the recovery of Portuguese sovereignty in Timor, which had been occupied by the Dutch and the Australians in 1941, and by Japan in 1942. In the second place, politically, if not a full assurance, at least a facilitation of the regime's survival after the war.[8]

In short, throughout the whole war, Portugal preserved the formal neutrality it had declared as early as September of 1939. It was, however, a relative neutrality that evolved with the conflict itself and varied according to the fluctuating interests of the belligerent powers and the balance of strengths within the peninsula. A neutrality that began as "geometric" and equidistant and ended up cooperating with the allied powers.

The above accounts for the nature of Portuguese neutrality, but not for the reason behind it. In other words, why neutrality?

Obviously, this was the diplomatic position that favoured to a greater degree the attainment of the objectives stated above. But the question is really what determined and made possible Portuguese neutrality.

Historiography has abundantly debated the issue of neutrality and its meaning in Salazar's foreign policy during the Second War. There are two main explanatory theses: a traditional thesis, conveyed by the regime's historiography, tends to explain Portuguese neutrality from the viewpoint of the "Statesman", that is, it credits the success of the neutrality policy to Salazar's political sense and diplomatic action.[9] The more recent theses of Portuguese and international academic historiography,[10] despite their differences, tend on the whole to downplay the action of the "Statesman" and to bring to the fore a number of objective factors of varying relevance: the strategic evolution of the conflict and the war objectives of the belligerent powers; the internal situation of Francoist Spain; and/or the very constraints of Portuguese foreign policy and military instrument.

There is no doubt that the political determination and diplomatic action of Salazar's government represented a major factor for the success of Portuguese neutrality. But it is also unarguable that Portuguese foreign policy would have known a dissimilar fate had the combination of circumstances in which it developed and which ultimately accounted for its success been different: in the first place, the fact that the neutralisation of the Iberian Peninsula served the interests of the belligerent powers, which found it convenient for the pursuance of their objectives and the military conduct of the war; in second place, the internal situation of Spain itself, whose impoverishment, inherited from the civil war and aggravated by the Allies' economic blockade, had destroyed all strategic stockpiles that might render it able to enter or conduct any sort of war.

Apart from and beyond the historiographical debate, this much is certain: neutrality favoured the position and the political objectives of the *Estado Novo*.

A neutral country with an authoritarian regime, Portugal had nothing to do with the international reconstruction and reordering of the post-war period. It was not invited to the Peace Conferences and found itself excluded from the international organisation that had ensued from the second post-war period: the United Nations.

Notwithstanding that exclusion, which would soon become isolation, Portuguese foreign policy managed to achieve three of the objectives laid out in the outset of the war: external security and peninsular duality; colonial integrity; and the political survival of the authoritarian regime.

What consequences would neutrality have on defence policy and the role of the military instrument in Portugal?

The avatars of the continental threat: Defence policy, strategy and the military instrument

The onset of the Second World War, in September of 1939, confirmed that which in Portugal was already a clear perception of the core threat to the external security of the State. Besides the "Spanish peril" in the peninsula, now there was, in the context of a European conflict, the possibility of an invasion of the Iberian Peninsula by Nazi Germany. The most likely prospect was still, and now *a fortiori*, that of the continental threat. Now, it is against the continental threat and, consequently, having the land border as top priority, that Portugal's defence policy between 1939 and 1945 will be drafted.

At the onset of the war, the state of the Armed Forces was such that it might have been one of the factors that determined Portuguese neutrality. The modernisation of the Armada, initiated in 1930, had been suspended in 1935 so that the Army could be granted priority. In truth, however, the situation wherein the Army found itself was not any better. Of the ambitious 1935–1936 modernisation plan, which in its more robust form stipulated the creation of ten divisions, no more than one modern division was actually in place. Not to mention the severe shortage in heavy weapons and the scarcity of ammunition.

Hence the Army faced the whole of the Second War with Santos Costa's 1937 reorganisation. However, a few changes – in doctrine and as regards equipment and armament – were introduced in the first half of the forties. In terms of doctrine, the most relevant change is the creation of the Institute of High Military Studies, intended to provide training to high commands and the Staff Corps. But there were other less far-reaching measures, such as the creation of new organic cadres and new rules of engagement for campaign units.[11] As to equipment and armament, the plan launched in 1935–1936 was still in force, although not without difficulties. It spanned for two distinct phases that corresponded to the different moments and avatars of Portuguese neutrality. Initially, the whole approach to rearmament was based on the assumption of British military cooperation with Portugal. However, the mounting tensions between the two countries, from 1937, in respect of the Spanish Civil War, which peaked in 1938, with the Munich crisis, precluded this solution. Portugal thus resorted to Germany in this first phase of the plan.[12] And in fact, from 1938, the rearmament of the Army relied on German war supplies. It was the case of the Mauser rifle (7.9 mm), the Parabellum pistol (9 mm), the Dreyse and Borsig machine guns (7.9 mm), the anti-aircraft machine gun (20 mm) and the howitzers built by Krupp and Rheinmetall. War material supplies also came from Italy, namely the Breda machine gun

(7.9 mm), mortars (8 cm) and mountain artillery (7.5 cm). Other supplies of lesser relevance for the rearmament of the Portuguese army also came from countries such as England, Denmark, Sweden and even Austria. In Portugal, the military industry, having revived since the Spanish Civil War the factories of Chelas and Trafaria, recalibrated to 7.9 mm some weapons that were still usable and focused on the manufacture of ammunition. This first phase of the Army rearmament lasted until 1943. The second phase ensued from the "cooperative neutrality" and the military "facilities" in the Azores granted to the old ally. War supplies, at this point, came from England: the Bren machine gun (7.7 mm), anti-tank cannons (5.7 cm), grenade-launchers (5 cm), artillery howitzers (8.8 and 14 cm), artillery pieces (11.4 cm) and coastal artillery (5.7, 15.2 and 23.4). Their purpose was essentially to defend the port of Lisbon. Also from this period is the Bren gun carrier, the Humber armoured car and the Valentine and Centauro fighting vehicles.[13] By the end of the war, in 1945, the no longer depleted arsenal was able to fully equip three infantry divisions and one armoured car battalion, and partially another two divisions. During this period and within the scope of the rearmament plan, the Military Aeronautics – which was still an auxiliary branch of the Army – registers a major leap forward, with the creation of thirteen Hurricane and Spitfire squadrons. The third phase of the rearmament, marked by supplies of American origin, would have to wait for the post-war period and Portugal's admission to the Atlantic Alliance.

The Armada had not been neglected during this period. The priority granted to the Army notwithstanding, the naval branch was recognised for its strategic relevance for the defence of the country confirmed during the Second War. The steps actually taken towards its modernisation were few, however. Between 1935, when the plan for naval rearmament was suspended, and 1939, when the war began, the Navy shifted its focus to reorganisation and training, with a view to adapting itself to the newly arrived equipment and weapons, and to the development of a new doctrine for its employment. On the other hand, it strove to create infrastructures for the naval bases in the Atlantic archipelagos and to develop a naval aeronautics. Both efforts met with little success. During the war, for the missions it carried out or might have come to carry out, those instruments would have proved strategically important. But in the absence of naval infrastructures in the archipelagos, of a naval aeronautics and with the rearmament plan on hold, the available resources were limited. On the other hand, the Navy's traditional strategic thinking and planning were themselves geared for surface threats, when in the course of the Second World War Portugal was chiefly exposed to submarine and aerial threats. Against these, the shortages were even bigger. Still, to the available naval resources were added twenty steam boats and fishing boats, to rehearse, as far as possible, a strategy of naval defence. These circumstances lasted from 1939 to 1943, when the Anglo-Portuguese agreement radically changes the situation of the Armada. With the stationing of British troops in the Lajes base, in 1943, a new phase of rearmament begins for the Portuguese Navy. With the military support of its British counterpart, the Armada then receives six patrol ships, anti-ship aircrafts and patrol aircrafts, and undertakes the modernisation of the Portuguese destroyers. Anti-aircraft artillery also was not forgotten. This phase in the Armada's rearmament, under British influence, extends beyond the war until 1948, with the arrival of three submarines, six launches, six patrol boats and two frigates. From 1939 to 1949, tonnage increased from 33,234 t to 51,265 t (54%). And from 1949, NATO will open a new chapter.[14]

Throughout the Second War, the budget for defence evolved slowly and gradually, peaking in 1941 only to decline once again to inferior numbers after the end of the conflict. In 1939, in the beginning of the war, it amounted to 33% of the total public expenditure and would peak in 1941 at 42.6%. It would remain high during 1942 and 1943, at 41%, and then gradually declined until reaching 22% in 1948. In accordance with the priority granted to the Army during this phase of the defence policy, the Navy's share on expenditure falls. In 1941, when war expenditure peaks, the Navy stood at a mere 19%, against the 81% of the Army, in the overall defence expenditure. On the eve of the Atlantic Alliance, in 1948, the ratio is once more balanced, with the Navy standing at 42% to the Army's 58%.

As regards civil-military relations, the period of the Second War confirmed and reinforced the process of political control of the Armed Forces by the *Estado Novo*. Initiated in 1933 and consolidated with the Spanish Civil War, the subordination of the military to political power found in the Second War and in Portuguese neutrality a highly favourable ground, that is, the permanent and plausible threat – never to be fulfilled – of impending war.[15] By the end of the conflict, with the Allies' victory on the horizon, a measure of restlessness returned to the military milieus as a way to exert pressure on the political power. Pressure on political decisions, such as the compliance with England's request in the "wolfram question", and on corporate issues, such as the increase in wages. The turmoil would continue in the post-war period in the shape of failed attempts to overthrow the regime, the last of which, in 1947, became known as the "Mealhada uprising". During the war, however, the atmosphere between the military and the political power was one of mutual understanding, which favoured the stability of the regime and Salazar's permanence in power. The longevity of the regime will be further established by Portugal's admission to the Atlantic Alliance.[16]

In this context of an evolving military instrument and against the strategic backdrop of the war, what are Portugal's defence plans, military missions and operational theatres?

Portugal's two major defence plans correspond to two strategic moments in war and in Portugal's stance towards the war: a first, during "geometric neutrality", between 1940 and 1941; and a second, during "cooperative neutrality", starting in 1943.

The first defence plan, as early as 1940, presumes the eventuality of the Nazis crossing the Pyrenees and invading the Iberian Peninsula. The Fall of France and the famous plan for Operation Felix made the Nazi progression a plausible threat. The Nazi plan was straightforward and stipulated that two German divisions should be stationed in Spanish Extremadura and in Andalusia, so that, should the need arise, they were able to march spectacularly onto Lisbon, which they would seize in a mere 48 hours. Such was the atmosphere of dreaded anticipation that hung in the air that, even before being informed of the Nazi plan, Portugal began to prepare for an eventual invasion. After discussing a defence plan with its old ally, the two countries reached an agreement in the spring of 1941. The plan was simple and focused on two core objectives: first, the Portuguese government should be transferred to the Azores, in the scope of a great naval operation; second, the defence of the mainland should be centred in Lisbon, along three distinct lines of defence. Naturally, the contingent available for the undertaking of such a plan was more than sparse. For that reason, Portugal requested England's aid in the form of a reinforcement consisting of six to eight divisions and two armoured brigades. England itself had no such forces available to lend, let alone with the promptness necessary to counter a swift German advance. Operation Felix was never launched and, therefore, Portugal's first defence plan never got off the paper. Even so, Portugal made, during this period, an enormous mobilization effort.

The second defence plan, in 1943, ensues directly from "cooperative neutrality" and of the granting of military facilities to England in the Azores archipelago, whose agreement provided for that very plan. Once again, the defence of the mainland was to focus on the Lisbon–Setúbal axis, now according to two lines of defence. Three Portuguese divisions were supposed to be allocated to this area, and joined by that many British divisions, one armoured and two infantry divisions, with the promptness requirement of 48 days. In the north of the country, the focus of defence being the city of Porto, a smaller Portuguese contingent was to be stationed, comprised of one mountain artillery group, one infantry battalion and one machine gun battalion. These were to be joined, with the promptness requirement of 29 days, by a British infantry brigade. In the south, the joint Portuguese–British net forces had the strategic goal of covering Portugal's two historical invasion lines, north and south of the Tagus River. The aerial defence of Lisbon and Porto was assigned to the Portuguese Aeronautics during daytime and to the RAF during night-time, which was also responsible for reinforcing daytime defence, if required. Naval defence was naturally the responsibility of the Navy.[17] As had been the case with the first plan, the second would also never be implemented.

Despite Portugal's diplomatic status of neutrality and its military non-involvement in the conflict, the Portuguese Armed Forces remain engaged in their missions: mobilisation and manoeuvres. Mobilisation, in a first instance, for the Atlantic archipelagos and the Empire. In 1939, the military personnel consisted of around 32,600 men and, with the mobilisation effort in 1942, around 11,600 men. From 1943, the contingent was reinforced, reaching 130,000 men, thus enabling the creation of three fully equipped divisions in the mainland, which were in still in place by the end of the war. A contingent of around 26,500 men was mobilised to the Azores, 3,400 to Madeira, 6,700 to Cape Verde and around 20,000 to Angola and Mozambique. The manoeuvres and military and civil protection exercises unfolded according to the approved defence plans, especially the major manoeuvres of October of 1943, in Pegões.[18]

The Armada, besides its traditional missions of troop transportation and military presence in the Empire, was assigned throughout the war with a whole new set of missions: the defence of ports and coastal navigation control in the mainland, protection of the archipelagos, escorting of ships and even, within the scope of the first defence plan, the preparation of the prospective major naval operation whose objective was to transfer the government to the Atlantic islands, in case of an invasion in the mainland.[19] From 1943, the Azores become top priority both for the Navy and the Air Force.

The operations carried out by the Armed Forces during the Second War did not take place in the mainland alone, but also in two other operational theatres: the Azores and Timor.

The strategic value of the Azores archipelago had always greatly interested both belligerents, especially the Western Allies, which developed plans for its use and, since 1943, actually implemented them. Whether for the defence of the mainland, or for its intrinsic strategic value, the archipelago represented a crucial – albeit extremely vulnerable – logistic support base. Both the Navy and, a while later, the Air

Force, are transferred *en masse* to the Azores, and faced tremendous operational difficulties ensuing from the lack of infrastructures in the military bases. Given these limitations and the low aero-naval capacity of the Portuguese Armed Forces, practically the whole military defence scheme is rehearsed from within the Army. Even the latter had to wait until 1942 for a clear definition of the concept of military defence of the Azores.[20] From 1943, its military function becomes clear with the need for a strategic articulation between the two shores of the Atlantic, and the arrival of the British and American troops in 1943–1944 will significantly alter the whole military panorama, especially in the aero-naval domain. Now, the post-war period and the Cold War confirmed the strategic value of the Azores both for national defence and, internationally, as a fulcrum for the strategic articulation between the two sides of the Atlantic.[21]

Occupied, since December of 1941, by Australia and the Dutch, Timor was the other operational theatre warranting Portugal's concern. Portugal called upon England for the replacement of the allied troops by Portuguese military forces, whose contingent had formed in Mozambique and later departed for Timor. However, the clock was ticking – diplomatic negotiations had dragged on for some time and the voyage of the Portuguese military contingent also took a while, and so, in February of 1942, when the Japanese invaded East Timor, the Portuguese were still in the middle of the Indian Ocean. It was not through military means that Portugal managed to recover Timor, but through diplomatic endeavours. In the framework of the negotiations for the installation of an American military base in Santa Maria in the Azores, in 1944, Portugal ensured in return, political sovereignty and the military reoccupation of Timor. This on the assumption of Portugal's participation in the Japanese surrender. But whereas the diplomatic manoeuvre was successful, the same cannot be said of the military efforts. A demurral in preparing a force for deployment and the distance to Timor would dictate, once again, the delay of the Portuguese. The Japanese capitulation takes place on the 22nd of September of 1945 without the military presence of the Portuguese, whose contingent would only arrive a week later.[22] Delayed or not, the diplomatic rapprochement and the military facilities granted to the United States indeed ensured, at least in the immediate post-war period, sovereignty over Timor and the integrity of the Empire.

Notwithstanding the two defence plans, the war had ended before the Portuguese Armed Forces had carried out a single military intervention. These plans, especially the one drawn up in 1943, were not devoid of importance, however. Quite the opposite, in fact. Not only, still within the context of the conflict, they framed military manoeuvres and civil defence exercises, but they also, and above all, left a significant mark in the military plan for the post-war period. Firstly, because the 1943 plan remained the base of rearmament plans for the Portuguese military instrument until the admission to the Atlantic Alliance. Secondly, because it left an imprint in Portuguese strategic thought almost until the fifties.

In 1947, Portugal submitted a proposal to England for the revision of the 1943 defence plan, accompanied by another rearmament plan. England, however, was engaged and concerned with a different strategic scenario, and its negative reply to Lisbon cleared the way for a new phase in Portugal's defence policy.

From 1945, the internal panorama had been radically altered, and the perception of threat had begun to change. In a bipolar world, the security threat looming over the Western world was no longer Nazi Germany, but rather the Soviet Union. For Portugal, the threat was still basically continental, but now emerged under a new avatar and possessed a two-fold nature: to the traditionally external military threat represented by the arrival of the Red Army to the Pyrenees, to Salazar's regime a new internal political and ideological threat was added: the "communist danger". This perception of a double threat will influence the strategic concept of the defence policy in the post-war period: Santos Costa's concept of the "Iberian Bloc", which guided the rapprochement between Portugal and Spain and drove the two Iberian dictatorships to set in motion a military cooperation system. Firstly, against the likelihood of a Soviet assault. According to the official strategic concept, the Red Army could only be contained in the Pyrenees. Thus, the Portuguese would contribute to the defence of the Iberian Peninsula with the deployment of three divisions. In the second place, against the danger of internal sedition and the external pressures that the wave of democratisation generated in the aftermath of the war, the two countries developed mechanisms for cooperation and mutual support in the defence of authoritarian regime. These were early inklings of the Cold War.

CHAPTER

6

The Cold War and NATO

The international standing of the *Estado Novo* in the second post-war period

Towards the end of the Second War, Portugal faced a period of international vulnerability. A neutral country, it did not take part in the diplomatic efforts aimed at restoring peace and reordering the new international system. Even though the authoritarian regime managed not to succumb to the wave of democratisation of the post-war period, Portugal experienced, between 1945 and 1949, a period of international marginalisation and uncertainty in its foreign policy.

The first tell-tale sign of its international marginalisation can be traced to April of 1945, when Portugal was not invited to the San Francisco Conference, whence the establishment of the new post-war international organisation proceeded. Unlike what had been the case with the League of Nations (LN), not only is Portugal not a founding member of the United Nations (UN), but also its first attempted application for membership, in 1946, is vetoed by the USSR at the Security Council.[1]

To the international isolation were added, during this period, a degree of political indecisiveness and of uncertainty in Portuguese foreign policy, reflecting incapacity to adapt to the new world order. Salazar appeared not to understand, or rather, accept the deep changes of the second post-war international scene.

To begin with, he did not accept the bipolar order and the emergence of the two superpowers and, in particular, England's decline and the rise of the United States as a great naval power in the Atlantic. Only the signing of the Lajes Agreement, in February of 1948, began to change that perception that the Washington Treaty and the foundation of the North Atlantic Treaty Organisation (NATO) came to confirm. Secondly, Salazar appeared not to realise the importance of the UN as the new world organisation. As with the League of Nations (LN), he also reacted against the return of multilateralism and the "international parliamentarism" that the UN stood for. Thirdly, the end of old Europe and of European powers as protagonists in the international scene. He failed to understand, moreover, that the rebuilding of Europe could no longer be tackled individually by each State, but had necessarily to be carried out in a framework of international cooperation. Finally, he failed to accept that the two new superpowers – although by virtue of different reasons – were both anti-colonialists. What is more, he neither understood nor accepted that the right of a people to self-determination was UN's cardinal principle, and that the contribution of many colonial peoples to the allied war effort made decolonisation an all the more inexorable process. Salazar's positioning regarding the second post-war world will have a decisive impact on the future evolution of Portuguese foreign policy's main vectors: inclusion in the Atlantic security system, reluctance in the face of European integration and uncompromising defence of the colonial empire.

The first signs of reluctance towards the new post-war reality became evident forthwith in Portugal's position regarding the Marshall Plan, in June of 1947. Salazar's inability to understand the international conjuncture, and his traditional mistrust of the Americans drove him to decline the North American offer, thus excluding Portugal from the first phase of the Marshall Plan in 1947–1948. The outbreak of the Cold War and the Portuguese financial and economic degradation eventually changed this initial position regarding American assistance, which led to Portugal's application to the second phase of the Marshall Plan, in 1948–1949.[2]

With the exacerbation of the Cold War, a new Soviet threat began to loom as the new framework of Atlantic security gradually took shape and, consequently, the American prominence in this context. In Portuguese foreign policy, and despite Salazar's well-known misgivings, the first sign of change and convergence with the new reality is the signing of the bilateral agreement on military cooperation between Portugal and the United States in February of 1948. Unlike the provisions of the 1943 agreement, and echoing the new strategic environment of the post-war period, the new agreement formalised the permanence of the North American air base in the Azores archipelago.[3]

Its reluctance towards the Marshall Plan, on the one hand, and the signing of the Lajes agreement, on the other, heralded the two future – and indeed contradictory – lines of Portugal' external orientation: self-exclusion from the process of European unification and integration within the Atlantic security system. If, to these orientations, are added the adamant refusal to decolonise and the defence, at all costs, of the colonial empire, it becomes clear what were the strategic options of Portuguese foreign policy until the end of Salazar's regime. It also follows that the two lines of Portugal's military presence in the second post-war international conflicts were: the Cold War, with its inclusion in the Atlantic security system and its admission to NATO, and the Decolonisation Wars, in three African theatres.

From neutrality to alignment: Portugal and the foundation of the Atlantic Alliance

With the onset of the Cold War, in 1947, and the Berlin Blockade, in 1948, the Western world realised that, not only was it facing a different threat, but the very framework of security had changed. In a world divided into two blocs, and Germany itself divided between them, it was evident that what threatened European security could no longer originate in Nazi Germany, but arose from the Soviet Union. It realised, moreover, that the framework of European security could not be strictly European anymore, but Euro-Atlantic. Now, it is within this context of international security that the United States decides to open negotiations with the Brussels Pact countries, with a view to the constitution of a security system for the North Atlantic: the future NATO.

These negotiations, originally involving the United States, Canada and the Brussels Pact countries (France, United Kingdom and the Benelux), were extended to other European countries, invited to join the foundation of the Atlantic Alliance. Successfully concluded, they led to the signing of the Washington Treaty in April of 1949. The foundations had been laid for the Atlantic Alliance and Portugal stood among the founding countries.

In this context, what were Portugal's position and role in the foundation of the Alliance? To answer this satisfactorily, three preliminary questions must be addressed.

The first question is: why did the negotiating powers decide to invite Portugal to join the Alliance? A neutral country during the Second World War, and home to an unrelenting authoritarian regime, Portugal had been excluded from the post-war reordering of the international scene – not to say marginalised – between 1945 and 1949. Why, then, the invitation to sign the Washington Treaty and become a founding member of the Atlantic Alliance?

The second question, conversely, is: why did Portugal accept the invitation? Why sign the treaty and join the pact when it meant such a radical change in Portuguese foreign policy as relinquishing neutrality, so hard to conquer during the Second War and so often invoked by the propaganda of the *Estado Novo* as Salazar's and the regime's crown of glory?

Finally, the question of consequences: what were the immediate and medium to long-run consequences of Portugal's admission to NATO for the country's evolution – in foreign policy, domestic policy and, obviously, defence policy and the reforms of the military instrument?

Why was Portugal invited? The answer is straightforward and relates to the geopolitical and geostrategic nature of the very grounds underlying the new Atlantic security system. Conceived against the Soviet threat, this system was based on two pillars: the first was the strategic potential of the United States, bringing up the rear and self-guarantee of the whole system; the second, forming the front line and immediate defence, was the ensemble of countries that signed the Brussels Pact in 1948. However, for the system to work properly, it was necessary to ensure not only the protection of the northern and southern flanks but, above all, the articulation between the front and the rear, that is, between the two pillars of the Alliance. This was the main reason why Portugal was invited to join. The territory of the Azores archipelago represented, furthermore, an added strategic value. Already an important military base during the First War, it proved decisive during the Second World War and vital in the course of the Cold War. Included in Pentagon's post-war military defence system, the Azores base had previously been the object of a bilateral agreement between Portugal and the United States in 1948.[4] With NATO, within a multilateral framework, its value was further increased and its role confirmed by the strategic concept of the Atlantic Alliance: in case of an attack coming from the East, the so-called "rapid reinforcement of Europe" relied inevitably on the Azores, this huge natural aircraft carrier in the middle of the Atlantic Ocean. The strategic value of the Azores archipelago prevailed over the authoritarian nature of Salazar's regime. The geopolitical and geostrategic factors were thus the true reasons for inviting Portugal.

But why did Portugal accept the invitation, when doing so entailed such a pronounced change in its international standing, which would lead it from neutrality to alignment?

Ever since the fifties and up to the present day, many have been the attempts by Portuguese historiography to answer this question. In the fifties, Henrique Martins de Carvalho, in what would prove to be the first and likely the most objective analysis of the problem, explained it from a geopolitical and geostrategic viewpoint: not only the Soviet threat rendered the Atlantic Alliance the only way out to Western security, but also the geostrategic situation of the Portuguese territory and the serious context of the Cold War compelled Portugal to join the Alliance, instead of simply signing a bilateral agreement with the United States.[5]

More recently, Franco Nogueira also addressed this issue. Focussing on Salazar's personality, and without overtly advocating for Portugal's admission in the Alliance, he comes forward with a thesis that may be outlined in two key ideas: in the first place, the importance of Portugal's role as founding member; in the second place, the ascription of paternity rights over the very idea of the Alliance to Salazar himself.[6]

Albano Nogueira also analysed the problem from the standpoint of Portuguese foreign policy, focussing essentially on Salazar's political positioning and diplomatic manoeuvring, and concluded that the true reason for accepting the invitation and, ultimately, for Portugal's accession to the Alliance, was as much ideological and moral – if not more – as political in nature. For Salazar, it was tantamount to an ethical imperative, a "crusade" against communism in the defence of Western and Christian civilisation.[7]

José Medeiros Ferreira also pored over the subject. From a foreign police perspective and considering especially the analysis of bilateral relations between Portugal and Spain, he ascribes paramount importance to this factor in Portugal's accession to the Alliance. Portugal's admission and Spain's exclusion might grant Portuguese foreign policy a preponderant place within the Iberian Peninsula and the role of privileged interlocutor outside the peninsular framework, and that would have been critical to Salazar's decision.[8]

Despite the differences in their theoretical assumptions and political positions, the theses mentioned above rely on a common tenet: they favour the "Statesman" to the detriment of the "deep forces", to use Renouvin and Duroselle's classical terminology. That is, they focus almost exclusively in Salazar's ideological, political and diplomatic positioning and attach less importance to the external environment and internal constraints.

Now, an understanding of the decision-making process must rely both on the attitude of the decision-maker and on the internal and external factors that influence the decision. And, despite the ideological, political and diplomatic motivations at play, Portugal's accession to the Alliance is ultimately linked to those constraints and, in particular, to the reduced leeway of Portuguese foreign policy.[9]

Salazar was aware, from an early stage, of the changes in the correlation of forces in the international scene and of the emergence of a new framework of world security, and had highlighted the advance of communism in Europe. It was in 1947, when giving a speech on current international events.[10]

For that reason, when, in 1948, at the UN's general Assembly, Mr Spaak drew the world's attention to the "Russian threat", Salazar presumably felt his concerns confirmed and his traditional anti-communism justified. However, he had no sympathy for the post-war multilateral diplomacy, let alone the democratic regimes that conducted it, especially the United States of America. It is thus unenthusiastically and even with demur that he receives the formal invitation to sign the North Atlantic Treaty. In the negotiations that preceded admission, between Portugal and the negotiating powers – England and the United States – Salazar insistently raised four objections to the text of the pact: in the first place, the allusion to the UN Charter, from which Portugal had been excluded, and the parliamentary democratic model, which the *Estado Novo* rejected; secondly, the refusal of any solution involving integration or supranationality, for fear that the true aim of the pact might be to obtain, in peacetime, the use of the Azores base – he feared the loss of sovereignty and did not want to lose control of it; thirdly, the case of Spain, which had been excluded from the pact and which Salazar, still under the influence of the concept of Iberian Bloc, deemed essential for Western defence; and, finally, the obligatory twenty-year commitment to the Treaty, a length of time he considered to be excessively long, fearing that Portugal might be involved in a new conflict after the neutrality conquered in the Second War.[11] Except for the guarantees concerning the strategic bases, and although their rejection was diplomatically justified, no other Portuguese request was granted. In fact, none of them seriously affected the great powers.[12]

As the scheduled date for the signing of the treaty approached, Lisbon began to be subject to diplomatic pressure from different countries. On one hand, Spain was pressuring Portugal not to join by invoking the Iberian Pact.[13] On the other, the United States and England urged Portugal to go ahead with the adhesion. Dean Acheson and Ernest Bevin themselves sent personal messages to Salazar on the eve of the council of the ministers which would decide the Portuguese position.[14] The pressure over Lisbon was mounting and there was little leeway. Internationally, Portugal's refusal might have a negative impact on countries in the process of pondering their own adhesion. It might open a breach in the West that would only benefit the Soviet Union. In a scenario of worldwide polarisation and exacerbation of the Cold War, a refusal by Portugal, with all its consequences, was too great a risk for Salazar to take before the political leaders and the public opinion of the Western world. Moreover, in the context of the peninsula, Portugal's refusal could be further interpreted as a sign of weakness and yielding before Francoist Spain.

In short, refusal meant running a significant risk, the leeway of Portuguese foreign policy was small and alternative options almost none. After three councils of the ministers,[15] the government divided and unenthusiastic, Salazar decided to join the pact. Less through conviction than necessity. On the 4th of April of 1949, Portugal signed the Washington Treaty and became a founding member of the Atlantic Alliance.

Finally, what would be the short and long-term consequences of Portugal's adhesion to the Atlantic Alliance at the various levels: that of domestic policy and that of *Estado Novo*'s foreign and defence policy?

Notwithstanding all the reservations that had delayed Portugal's final decision, its admission in NATO represented paradoxically a victory for Salazar. A victory *malgré lui*, but still a victory.

To begin with, a victory of foreign policy. In the framework of the peninsula, despite Salazar's insistence for Spain to be admitted in the Alliance, the refusal by the Western powers proved beneficial for Portuguese foreign policy. The coincidence of Portugal's admission and Spain's exclusion unmistakably reinforced Portugal's position in the peninsula and establish Lisbon as chief interlocutor in the extra-peninsular relations, especially with the United States.[16] Globally, Portugal's adhesion to the Alliance came to confirm the permanence of the historical vectors of Portuguese foreign policy, that is, the Atlantic option and the alliance with maritime power. However, there was a fundamental change in its pole of reference which was the prevalent naval power in the Atlantic: the decline of the Anglo-Portuguese Alliance and the emergence of a new axis on the other side of the Atlantic – the United States of America. That change would constitute the Portuguese answer to the international scene of the second post-war period.

Secondly, insofar as this was a military pact, the consequences of Portugal's admission in the Alliance were naturally felt in the military institution itself and in defence policy. Despite a few clashes, within the military hierarchy, over Portugal's military defence concept, the military institution on the whole supported the adhesion to NATO. They hoped that the ensuing military aid from the United States might make a vital contribution to the professional modernisation and the military re-equipment of the Portuguese Armed Forces, which indeed occurred throughout the fifties under a "mutual aid plan". On the other hand, this period of professionalisation and modernisation of the Armed Forces was accompanied by a withdrawal of the institution into itself and an absence of military intervention in political life, which further worked in Salazar's favour.

Thirdly, at the level of domestic policy. If it is a fact that Portugal's admission in NATO divided the regime, it is no less true that it caused even deeper and more enduring dissension within the opposition: the communists, siding with the Soviet positions, opposed radically and regarded the Alliance as the military instrument of North American imperialism;[17] the democratic opposition, on the other hand, seeing in the pact an embodiment of the principles of the United Nations Charter and of democracy, supported enthusiastically the Portuguese adhesion, hoping that it might entail an implicit condemnation of authoritarianism and a chance, even if remote, of the democratisation of the regime[18].

The outcome would be the exact opposite: the acceptance of the Portuguese dictatorship in the midst of the Alliance, side by side with Western democracies, not only conferred to the regime a measure of international legitimacy, but also aggravated the divisions among the opposition and cast it into one of its more enduring crises, extending throughout the whole decade of the fifties, until General Humberto Delgado's presidential candidacy.[19]

Hence Portugal's adhesion to NATO, in 1949, launched one of the most stable periods of the history of *Estado Novo*, in terms of both foreign and domestic policy. Only Portugal's admission in the UN, in 1955, and the start of international contestation against Salazar's colonial policy eventually changed the panorama and closed this period.

The Atlantic option and Portugal's presence in the Alliance will remain, however, basic references in the foreign orientation of the Portuguese State until the end of the authoritarian regime and will even be confirmed with the transition to democracy.

Portugal and NATO: Defence policy and the reform of the military instrument

From the moment when the Alliance was ratified and Portugal signed its adhesion to it, except for the brief period that followed the revolution of April of 1974, NATO completely vanished as an issue of domestic policy in Portugal. Its consequences were felt, however, throughout the whole period, at the level of foreign policy and defence policy.

For Portuguese foreign policy, Portugal's entry into NATO meant not only the emergence of a new preferential ally – the United States – as a qualitative change in the bilateral relation between the two countries, especially in the field of defence. From 1949, although the military cooperation and the so-called facilities granted by Portugal to the United States ensued from bilateral agreements, their relationship would of necessity acquire a multilateral dimension and be seen as Portugal's contribution to the Alliance's defensive system. As such, to grant or to restrain military facilities was no longer a strictly bilateral problem, but a matter for NATO.

From this moment onwards, and inasmuch as the United States represent the true military mainstay of the Alliance, diplomatic relations between Portugal and the United States, especially involving cooperation in terms of defence, become the fulcrum of Portugal's relationship with NATO. In other words, for Portugal, the transatlantic relationship acquired two dimensions: a bilateral one, involving the United States, and a multilateral one, involving NATO.

In that sense, the fifties are marked by a good bilateral relationship and, within that context, by the signing of agreements on military cooperation. Following the agreement on the Azores base, in 1948, and the Mutual Defense Assistance Act, in 1949, always within the framework of the Atlantic Alliance, Portugal and the United States sign an agreement on mutual assistance in January of 1951, and, in September of the same year, a defence agreement that extended North American rights to facilities in the Azores military base until 1956. In November of 1957, the signing of a supplementary defence agreement renewed the agreement, extending it until December of 1962.[20]

As to defence policy, Portugal's integration in the military system of the Alliance would be rife with consequences. To begin with, consequences at the level of Portugal's strategic concept. Portugal's admission in NATO and the strategic role of the Portuguese territory in the framework of Western security was the object of debate and divergence of opinion among Portuguese strategists. The opposing strategic conceptions for Portugal's defence were then materialised in a controversy between General Santos Costa, who spoke up for the official thesis of the "Iberian Bloc", and General Raul Esteves, who advocated the strategic differentiation of the Portuguese territory vis-à-vis Spain. To Santos Costa, Portugal was Europe's rear-end defence. To Raul Esteves, Portugal should be US's forward defence. According to the former, defence should be carried out in the Pyrenees, which presupposed an eminently terrestrial conception. The latter argued for an Atlantic defence, which valued the territory of the Azores and pointed to an essentially aero-naval conception of defence. In the immediate term, defence policy continued to be grounded in the official thesis of the "Iberian Bloc". Time and NATO eventually proved Raul Esteves right. Despite the divided opinions on matters of strategy, among the military elite, the military institution as a whole supported Portugal's adhesion to the Alliance. They hoped that the North American military aid would contribute decisively to the modernisation of the Armed Forces and, if possible, to secure a few corporate advantages. It is indeed what happened throughout the fifties, not only in the Armed Forces, but also in the very conception of national defence. NATO would represent, from this point onwards, the pivotal change in the *Estado Novo*'s defence policy.

In accordance with the new global and integrated defence concept, the first change soon took place in the institutional organic structure of what would thenceforth be called "National Defence". In 1950, the office of Minister of National Defence is created, which, however, did not correspond to an autonomous ministry. The Minister of Defence is integrated in the Presidency of the Council of Ministers, and is essentially responsible for the strategic orientation of defence policy and the coordination of the three branches of the Armed Forces. The War Ministry, a traditional title that no longer reflected the new strategic concept, was replaced by the Army Ministry. The Navy Ministry is preserved and, given the growing strategic importance of the Air Force, since the Second World War, a Sub-secretariat of State for Aeronautics is created. It was the beginning of the Air Force as autonomous branch of the Armed Forces, which came into force in 1952. In the scope of the same reform, the General Secretariat for National Defence was also created. It was designed as a consultation and advisory body of the Defence Minister, at the head of which appeared for the first time the office of Chief of the General Staff of the Armed Forces. This decree was reinforced, in 1952, by the Basic Law of National Defence and, in 1956, by the Law of the General Organisation of the Nation during Wartime, enhancing the autonomy of the military institution regarding the political power and creating new bodies, such as the Chief of Staff of the Armed Forces and the Supreme Military Council.[21]

These reforms had a primary objective: to close the gap between the institutional structures of National Defence and the Portuguese Armed Forces and those of the other countries within the Atlantic Alliance.

They all aimed at adapting the Portuguese military instrument to the higher standards of the Alliance and, ultimately, at complying with the military commitments made within the framework of NATO. These commitments basically involved two types of missions: with regard to the Army, the creation of an autonomous division allocated to NATOS's missions in mainland Europe; with regard to the Navy and the Air Force, their engagement in missions of anti-submarine warfare and patrolling in the North Atlantic within NATO's commands.[22]

The first years of the Portuguese participation in the Alliance were still ruled by the strategic concept of the "Iberian Bloc", and with help from NATO, the Portuguese military commanders hoped to be able to complete the third phase of Santos Costa's old rearmament plan, which envisaged a gigantic Army with fifteen divisions. Gradually, throughout the fifties, NATO's doctrine changed the strategic conceptions still in force in Portugal: in the first place, it showed that, more effective than an outdated massive army was a force smaller in size yet highly trained and equipped; secondly, in line with what General Raul Esteves had already claimed, it highlighted the aero-naval dimension of the defence policy. These two trends will have a profound impact at the level of training and professional specialisation of the military cadres, and also at the level of the re-equipment of the Armed Forces. The whole modernisation process will be developed in close liaison and according to the models of North American Armed Forces: in the fields of organisation, doctrine and armament.

Between 1950 and 1956, the Army eventually abandoned the old goal of the fifteen divisions. In line with NATO's new doctrine, the new goal cut them down to only three, but focused mainly on the creation of a modern division after the fashion of NATO. In 1955, the first units were created – the Divisional Group of Fighting Vehicles and the Divisional Company of Equipment Maintenance. Finally, in 1957, with 18,000 men, the division was complete and set up its headquarters in Santa Margarida. It was called the 3rd Division or Nun'Álvares Division.

The Territorial Organisation of the Army, intact since the 1937 reform, was also revised in 1958. There was no longer a distinction between metropolitan army and colonial army, and the organisation encompassed the entire territory, divided into five military regions and seven independent territorial commands: the Military Government of Lisbon and the 21st and 2nd Military Regions in the mainland, with headquarters in Porto and Évora; two independent territorial commands, in the Azores and Madeira; the 3rd and 4th Military Regions at the colonies, as well as five independent territorial commands. This territorial organisation was altered on two other occasions: in 1962, with the creation of one more military region, in the mainland, headquartered in Tomar; and another, in 1970, based in Coimbra.[23]

The reforms undertaken within the Navy were no less significant and aimed at organisational, technical and tactical modernisation. At the organisational level, and despite the preservation of the Navy Ministry, its entire organic structure was revised, as well as that of the high command and the operational forces. Technically, novel systems of anti-submarine warfare and mine-sweeping were developed, as well as new command and radar systems. Tactically, new procedures were put in place in accordance with NATO. As to the quantitative evolution of naval forces, the Armada recorded, throughout the fifties, a growth of around 25%. In 1949, it possessed 57 ships amounting to a tonnage of 51,265 t and, in 1961, 71 ships with an overall tonnage of 64,332 t.[24] During this period, however, the Navy loses its Naval Aeronautics component, with the autonomisation of the Air Force as an independent branch.

In fact, in 1952, the Air Force became an autonomous branch of the Armed Forces and began to develop the first jet fighter squadrons and the first air control system. In the mid-fifties, the Portuguese Air Force has already in operation two F-84G squadrons, one PV-2 reconnaissance squadrons, two Helldiver anti-submarine squadrons, and, in 1958, F-86E, were added. It also had two FPS-28 radar stations.[25]

NATO also proved crucial for the renewal of Portugal's armament. In the framework of the Alliance, Portugal and the United States signed, in January of 1951, a "Mutual Defence Assistance Agreement". It was under this agreement that, in the course of the fifties, the whole process of modernisation of Portuguese military equipment and armament was carried out. It focused on two areas: the building of infra-structures and the reception of war material. A new cycle was thus launched for the rearmament of the Portuguese Armed Forces, this time under North American influence.

The investment in infrastructures was directed mainly at the construction of air and naval bases. As to armament, the Army received 12.7 mm Browning machine guns, 8.9 cm Bazooka grenade and rocket launchers, 5, 7, 7,5 and 10.6 cm recoilless rifles, 60 mm mortars, Ford and White armoured cars, Foz auto machine guns and around fifty M5A1, M24, Sherman and Patton combat vehicles.

The strategic priority having shifted to air and sea, the Armada was not forgotten and received, in its turn, fourteen minesweepers, nine patrol ships and three frigates. Portugal also acquired, for the Armada, in the marketplace for second-hand armament, three more submarines, six minesweepers, two frigates and two corvettes. And, among other weapons, the Napalm bomb.

The Air Force received, in a first phase, in 1953, 25 F-84G jet fighters, and, in a second phase, in 1958, 40 F-86F.[26]

Despite the ever-growing dependence on foreign technology – especially with respect to the more sophisticated armament – in the fifties, a process of development of the Portuguese military-industrial complex was set in motion which was further reinforced during the colonial war, and aimed essentially at the sectors of light weaponry, ammunition and the aeronautic industry.

During this period, military expenditure vis-à-vis the total public expenditure varied between 23.3% in 1949 and 26.1% in 1960, having reached a peak at 31.3% in 1954 and its lowest at 17.4% in the following year. Of the three branches of the Armed Forces, the Army remained always a heavier burden on the budget.

On the other hand, in parallel with the technical modernisation in infrastructures and armament, the Portuguese Armed Forces, in liaison with the North American Armed Forces, launched another process – as important or perhaps even more – destined at the recycling, training and professional specialisation of the Portuguese military. The fifties saw a complete renewal of doctrine, methods and procedures in the military institution.

Finally, throughout this period, NATO's influence on defence policy had important repercussions on the relationship between the military institution and political power, that is, between the military and the regime. On one hand, the professional training of the Portuguese military, in connection with the North American Armed Forces, gave rise to the emergence of modernising military elite, a "generation NATO" which gradually opposed Santos Costa's conservative leadership, which it finally replaced in 1958. On the other hand, the modernisation process of the Armed Forces was accompanied by a concomitant process of professionalisation, which led to a withdrawal of the institution into itself and an absence of the military in political life, all throughout the fifties. In fact, whereas the last great attempted military coup against the regime had been carried out in 1946 – the "Mealhada revolt" – after Portugal's adhesion to NATO, the next one will not be undertaken until 1961, with the "Botelho Moniz coup".[27] The latter, however, was invested with an altogether different political meaning – it was indicative of a new problem and marked the opening of a new phase ruled by the colonial question.

In short, the fifties were marked, in terms of foreign policy, by a smooth bilateral relationship with the United States, sealed by the signing of defence treaties and the mutual assistance plan, and in terms of defence policy, in the multilateral framework of NATO, by the engagement in the military commitments taken with the Alliance.

The sixties, on the contrary, were marked by the decolonisation conflicts and the difficult relations between Portugal and the United States, by the failed bilateral negotiations over defence agreements, and, multilaterally, by the non-fulfilment of the military commitments with NATO.

Portugal's admission in the UN, in 1955, opened a new period for Portuguese foreign policy, marked by international contestation directed at the colonial policy of the *Estado Novo*.[28] First, at the ideological level, the clash with the anti-colonialism prevalent in the General Assembly of the UN; then, at the political level, the confrontation with the liberation movements; and, finally, at the military level, the question of India and the outset of the decolonisation conflicts in three African theatres. From 1961, Portugal abandoned its engagement with the Atlantic and suspended, almost completely, the military commitments with NATO. The colonial question had become the Portuguese foreign policy's almost exclusive concern.

On the other hand, from a North American point of view, the year of 1961 also coincided with the rise to power of the Kennedy administration and the beginning of its interventionist policy at the international level. The influence of Africanists in the Department of State was heavy and pressure for Salazar to reverse his colonial policy did not take long to arrive. It was substantiated by the so-called Sakwa Plan.[29]

Now, the collision of these two positions – Salazar's intransigence with respect to his colonial policy and Kennedy's liberal interventionism – had immediate effects on the bilateral relations between Portugal and the United States, especially on what concerned bilateral defence cooperation.

The 1957 supplementary defence agreement had a five-year duration, thus expiring in December of 1962. When, near the middle of the year, the American ambassador to Lisbon requested the Portuguese government the reopening of negotiations with a view to renewing the agreement, Salazar had the firm intention of not negotiating. Not even the US's favourable vote at the UN, nor the signs of a greater tolerance towards the colonial question, nor the visit of the North American Under Secretary of State George Ball to Lisbon, in a last attempt to secure a deal, managed to change Salazar's mind.[30] Since early 1962, in the scope of a NATO meeting, the Secretary of State Dean Rusk had conveyed to the Portuguese Minister of Foreign Affairs once and again his apprehension regarding the operational problems that might be brought forth by such a situation. In the face of this, the Portuguese government showed itself available to find a solution for the problem, but not without making its position very clear. It issued a note wherein it authorised the stationing and the use of the Azores base, but, it must be stressed, "only *de facto* and during the negotiations".[31] To all practical purposes, the agreement was extended until 1964. Unable to drive away the North Americans, Salazar made it clear that it was nonetheless a *de facto*, and not a *de jure* arrangement, thus apposing on the whole thing the unequivocal seal of precariousness. Precariousness indeed marked the times ahead, since an aggravation of the colonial question, as well as the deteriorating relations between Portugal and the United States in the last years of Salazar's government precluded any possibility of successful negotiation.

Militarily, and from the point of view of defence policy, the sixties were largely marked by a neglect of the commitments with NATO.

In 1961, the Defence Minister General Botelho Moniz, whose dealings with the Kennedy administration are known today,[32] led a military coup intended precisely to reverse Salazar's colonial policy and to seek a non-military solution, that is, a negotiated political solution for the decolonisation conflicts. As the coup failed, all the political priority was directed towards the empire and all the Portuguese military effort to the colonial theatre.[33]

Completely immersed in the ongoing war in Africa, the Portuguese government was forced to substantially reduce its commitments with NATO.

In the first place, at the Army level, in which the greatest change was felt. The Army's independent division – to which the US administration had ceased to provide military support since 1962, with the explicit purpose of preventing war material granted to the Portuguese Army from being used in the African war theatres – was dismantled.[34]

As to the Navy and the Air Force, despite the reduced engagement in the commitments with NATO, changes were not so radical. Patrolling missions continued and, since the creation of STANAVFORLAND, in 1967, the Atlantic naval force relied on the constant (four to five months a year) participation of a Portuguese frigate.

At the strategic-military level, during the fifties, the Atlantic Alliance underwent another development that had an important impact in Portugal: the redefinition of the SACLAND operational theatre and of its command structure. To the two operation zones already defined – WESTLAND and EASTLAND – a third was added: IBERLAND. And although the decision to create the IBERLAND was taken in 1958, divergences between France and England over the command of the operational area made its establishment drag on for years. Only in 1966, coinciding with the French withdrawal from NATO's integrated military structure with all its commands,[35] was a compromise solution reached: the command of IBERLAND would be based on Portuguese territory and its command would be assigned to a North American officer. In February of 1967, the IBERLAND's headquarters were provisionally set up in Sintra, wherefrom it moved, in October of 1971, to the new premises that to this day it occupies in Oeiras.[36]

During the first half of the seventies, and until the transition to democracy, there were no noteworthy changes in defence policy and Portugal continued to be little engaged in its commitments with NATO, since the conflicts in the three African theatres persisted until 1974.

In foreign policy, on the contrary, there were notable changes. Advances in the bilateral relations between Portugal and the United States allowed for a solution to the pending problem of the Azores base.

With Salazar's exit from the political scene in 1968, and Marcello Caetano's rise to power, a revision of foreign policy was initiated, particularly in what concerned relations with the United States. Unlike Salazar, Caetano avoided political tension and ideological confrontation, and sought to cooperate as much as possible with the United States government. Portugal's new positioning was favoured by the rise to power, at roughly the same time, of the Nixon administration, which was itself more benevolent to Lisbon's colonial policy. In light of this, as early as 1969, the Portuguese government informed Washington of its interest in solving the question of the Azores, and proposed to start negotiations to that purpose.

The principles guiding these negotiations were altogether different from those prevalent in Salazar's time. In the first place, Portugal was not asking the US government for any political assurances – which the latter, in any case, was not in a position to give – which prevented political frictions. Secondly, and due to the fact that Portugal was under an embargo on arms supply, Lisbon was forced to negotiate, not for military compensations, but rather for economic-financial ones. This more realistic and pragmatic approach allowed the impasse to be overcome and a new phase for Portuguese–US bilateral relations to begin.

The negotiations were successful and the agreement was signed in Brussels, after a NATO meeting, in December of 1971. The agreement was essentially based on the following conditions: Portugal would extend the facilities in the Azores until February of 1974; and, in return, the United States would offer Portugal a vast programme of economic assistance. This economic assistance programme consisted basically of the following four points: first, a 15 million dollar loan for 1973 and 1974, redeemable in fifteen years at an interest rate of 4.5%; second, the provision of a 400 million dollar credit line for projects in the scope of economic development; third, the donation of one million dollars for development projects in the education sector; fourth and finally, the offer of a hydrographic ship.[37]

It represented a shift in paradigm. If, between 1948 and 1971, the bilateral defence agreements had been based on military compensation, from 1971, a new phase began in which the renewal of facilities was tied to economic-financial compensation. This trend extended beyond the authoritarian regime, into the democratic transition and consolidation, and lasted until the 1989 agreement. From the nineties, the end of the Cold War, Portugal's accession to the European Community, the emergence of other countries requesting the same assistance, the US budget restrictions, and, above all, the question of burden-sharing within NATO, altered significantly the tenets of Portugal's traditional basis for negotiation, that is, the facilities/compensation binomial.[38]

In the last years of the *Estado Novo*, two new events brought about an improvement in relations between the two countries and a strengthening of Portugal's negotiating position.

First, in October of 1971, when, in the General Assembly of the UN, the question was raised of Chinese representation at the Security Council, that is, of which of the Chinese republics was to have a seat at the Council, Portugal voted in favour of North American interests.

Second, and this would be the crucial element, Portugal's authorisation for the use of the Lajes Base, in the Azores, during the Yom Kippur War, in 1973.[39] Portugal's position had such a positive effect on bilateral relations that, despite the embargo on arms supply still in force, Henry Kissinger's Secretary of State endeavoured to secure the supply of missiles required by the Armed Forces for the conduction of war in Africa, especially in Guinea. Missiles that naturally were supposed to arrive in Portugal "through devious ways".[40] Such was then the atmosphere of bilateral relations with the US and multilateral relations with NATO when, in Portugal, the 25th of April of 1974 takes place.

However, Portuguese foreign policy and the whole international standing of the *Estado Novo* during the sixties and the first half of the seventies had been infused by one other problem, without which neither of the above can be understood: the colonial question and the decolonisation conflicts in the last African empire.

CHAPTER 7

The End of Empire and the Decolonisation Wars

The obdurate defence of the preservation of the empire by the colonial policy of the *Estado Novo*, in counter-cycle to all international trends, led Portugal, between 1961 and 1974, to three military conflicts in three different operational theatres in the African continent: Angola, Guinea and Mozambique. Overseas wars for some, liberation wars for others, colonial wars for others still, these conflicts constitute, at bottom, an integral part and the final phase of the decolonisation processes. Hence, it is as decolonisation wars that these conflicts must be understood.

Decolonisation is not limited to the final, formal act whereby the sovereignty held by the colonial power is transferred to the new independent state. Contrary to the traditional thesis that reduced decolonisation to a simple, short-lived event of an essentially juridical and political nature, current theories produced by historiography and social sciences view decolonisation as a complex long-term process involving the whole society in a coloniser/colonised dialectic. According to some, colonisation and decolonisation are two sides of the same coin, opposed dynamics of a single phenomenon.[1] According to them, decolonisation is, ultimately, resistance to colonisation and traverses different moments in its historical manifestations and different processes depending on the colonial model. In its modern sense, anti-colonialism emerges between the two wars, affirms itself in the second post-war and triumphs in the successive decolonisation waves between the fifties and the seventies, with the entry of the new independent states into the international scene. In that process, anticolonial movements came a long way, since the more or less spontaneous revolts against the injustices of the colonial status and the training and the growing awareness of anticolonial elites, to the political constitution of pro-independence parties. In this final phase of the decolonisation process, the struggle for self-determination and for the independence of colonised peoples goes through a period of political and diplomatic affirmation of the pro-independence parties and, in many cases, armed struggle and military conflict.

The last European colonial empire, the Portuguese case was no exception. Therefore, in order to overcome the traditional dichotomy, politically and ideologically marked in Portuguese historiography between two antagonistic designations, we shall opt for the use of this theoretically more precise concept. The expressions "African war" or "overseas war", on one hand, conveyed by the political discourse of the authoritarian regime, tend to elude the colonial nature of the conflicts, and, on the other, the designations "colonial wars" or "national liberation wars", relayed by Portuguese opposition or by the liberation movements, holds a negative or heroic connotation. We therefore shall use the concept of "decolonisation wars".

The international isolation of the *Estado Novo*: The international scene, Portuguese foreign policy and the colonial question

Portugal's adhesion to the Atlantic Alliance, in 1949, launched one of the most favourable periods of the *Estado Novo* foreign policy, which extended throughout the fifties. After its position of neutrality during the Second War, Portugal embraced alignment and entered the Atlantic security system. Included in the Western Bloc, the *Estado Novo* earned a measure of international legitimacy, which was soon made evident by its sound bilateral relations with the United States and the other Western allies, as well as by the commitment to its obligations within the multilateral framework of NATO, which translated positively in the country's international standing.

Portugal's admission in the UN, in 1955 – expected to be the crowning of that favourable international standing – represented, on the contrary, the first sign of change and the onset of the problem that would overhang the country's foreign policy and international standing during the sixties: contestation aimed at Portuguese colonial policy, at a political and diplomatic level, and the outbreak of the decolonisation conflicts, at a military level.

In fact, in December of 1955, together with several other countries – Spain among them – Portugal is admitted as a member of the UN. From that moment, the international positioning of the two Iberian countries begins to move in opposite ways. Spain's admission, following the signing of defence agreements with the United States,[2] meant the beginning of the end of the international isolation of

Franco's regime. For Portugal, on the contrary, it meant the end of complacency towards the *Estado Novo* and the launching of an international campaign against the regime's colonial policy.

Indeed, except for the possibility of a multilateral foreign policy, the immediate and most relevant consequence of Portugal's admission in the UN was the start of a confrontation with the anticolonial spirit of the non-aligned Afro-Asian movement already prevailing in the General Assembly of the United Nations.

In fact, between 1956 and 1961, a concern began to emerge in Portuguese foreign policy that would become almost exclusive during the sixties and the first half of the seventies. Throughout this period, Portuguese diplomacy was busy seeking external support for Portugal's overseas policy. In this context and to this purpose, from 1957, a cycle of state visits was launched, starting with the President of Brazil, Jucelino Kubichek, in an effort to establish a Portuguese–Brazilian community, followed by the President of Pakistan, Queen Elizabeth II of England, President Sukarno from Indonesia and the Emperor of Ethiopia, Haile Selassie. In 1960, the same policy remained in force, with the visits of the presidents of Peru, Nepal and Thailand, culminating in the visit of the Secretary General of the United Nations, Dag Hammarskjöld, and, especially, of President Eisenhower.

Despite this diplomatic offensive, however, the international atmosphere surrounding Portuguese colonial policy did not change. On the contrary, it deteriorated, and after the ideological confrontation at the General Assembly of the UN, Portugal was facing diplomatic and military conflict. First, with the Indian Union, and then, with the African liberation movements.

In 1950, the Indian Union issued a statement formally claiming Indian sovereignty over the territories of Goa, Daman and Diu and asking the Portuguese government to open negotiations to that purpose.

Salazar found himself, for the first time, faced with the problem of decolonisation. On principle, he had never accepted "the right of peoples to make their own choices and decide their own fate", and, for that reason, he could not accept its political corollaries: the self-determination and the independence of colonised peoples. But now he was facing, not the theoretical principle, but a real prospect. The territories of Goa, Daman and Diu had little economic or even strategic significance, but, at that particular moment, were invested of huge symbolic value: should he accept Indian sovereignty over those territories, on what grounds would he defend Portuguese sovereignty over the remaining overseas territories? His goal was therefore to maintain colonial integrity at all costs. The strategy of Portuguese foreign policy sought to deny political and diplomatic options to the Indian Union, in the assumption that Nehru, in his self-professed pacifism, would not resort to the military option. Should he do so, Salazar thought, Portugal could count on the support of its closest allies. A double miscalculation: in December of 1961, the Indian Union invaded and occupied Goa, Daman and Diu.[3] And no one supported Portugal's position – not the former colony, Brazil and neither the old ally, England. The latter had already cautioned, through MacMillan, that the colonial question was excluded from the Anglo-Portuguese Alliance.[4] Francoist Spain, in its turn, had established in 1956 relations with Nehru's India. It was a first sign of a breach in the peninsular friendship. [5] None of the NATO partners supported Portugal. The international community watched the unfolding events with aloofness and the General Assembly of the UN with a degree of satisfaction.

Between 1961 and 1974, the international isolation of the *Estado Novo* became more pronounced and the highest peak of the well-known "proudly on our own" policy was reached.

Multilaterally, the main stage of that isolation and, subsequently, hostility, was the United Nations. NATO, on the contrary, Portugal's non-compliance with its military commitments and the political discomfort of some of its allies notwithstanding, was nonetheless the sole diplomatic instrument that assuaged Lisbon's international isolation.

Bilaterally, the international relations of the *Estado Novo*, especially with its Western Bloc allies, were shadowed by an overall atmosphere of ill-feeling and ambiguousness. The greatest strain was with the United States, which, from the Kennedy administration, exerted overt pressure in condemnation of Salazar's colonial policy, and even essayed a number of political and diplomatic solutions at bringing about the decolonisation of Portuguese Africa. Relations with its European allies were less tense and more lenient, as a recent past of attempted resistance to decolonisation, until the sixties, was still a fresh a memory for a number of them, such as France. Even with the United States, the tougher and almost confrontational position of the Kennedy administration gave way to Nixon's more tolerant, almost cooperative approach.[6]

After the loss of India, the problem worsened and spread to Africa. In July of 961, the President of Dahomey orders the retreat of the Portuguese from the Fort of São João Baptista de Ajudá. This retreat was little more than symbolic since, except for the flag, the Portuguese presence amounted to no more than a clerk and two guards. Further ahead in 1961, however, war broke out in Angola and soon Portugal was involved in three decolonisation conflicts in three different operational theatres: Angola, Guinea and Mozambique.

The year of 1961 will indeed prove to be a key year for the colonial question. As early as January, Captain Henrique Galvão hijacked and diverted the cruise liner Santa Maria in the Caribbean Sea and remained in the high seas for twelve days, broadcasting anti-Salazarist messages and generating an atmosphere of speculation as to the ship's destination and his own objectives. The coincidence with the first revolts in Angola spiked conjectures that he might be headed towards Luanda with the purpose of establishing a government in exile. The United States did not shrink from voicing their diplomatic support to Lisbon, yet never openly condemned the occurrence as an act of piracy.

In April, a few weeks after the outbreak of conflicts in Angola, a group of military led by the Defence Minister, General Botelho Moniz, with the knowledge and consent of the US administration, essays a coup against Salazar – the 1961 *Abrilada*.[7] The coup's objective was precisely to remove Salazar and change the colonial policy. To discard the military solution and to embrace a political and diplomatic solution for decolonisation.

When the coup failed, any chance of a negotiated solution for the colonial question was cut short. The strategic goal of the *Estado Novo* was then obdurately directed towards the defence of the "colonial status quo".

On the last day of 1961 and the 1st of January of the following year, still one other attempted coup by a number of rebellious officers from the Beja headquarters was botched.

In all these events, rightfully or not, Salazar saw the "hand" of the United States of America. To be sure, in April of 1961, Washington voted, for the first time at the United Nations, on the side of the Soviet Union and against Portugal, in the question of Angola. Months later, it banned the supply of arms to Portugal that might be destined for war in the colonies. And, about the same time, it began to support the African liberation movements, especially Holden Roberto's UPA, in Angola, and, later on, Eduardo Mondlane's FRELIMO in Mozambique.

These were the first signs of the new African policy of the Kennedy administration, which multiplied and translated into various forms of diplomatic pressure over Lisbon, including the proposal of different plans, with a view to persuade it to change its colonial policy and seek a negotiated solution to the process of decolonisation.[8]

The first plan proposed by the US administration, in March of 1962, was the so-called Sakwa Plan. Drawn by the CIA, the Sakwa

Plan was based on the assumption that Portugal was bound to fight a long and unsuccessful guerrilla war, which would predictably represent a political nuisance for NATO and, worse still, the alignment of the new African countries with the Soviet bloc. In the midst of the Cold War, this was no minor issue. Between Salazar's integrationism and the uncompromising independence advocated by the Afro-Asian movement, the North American proposal proffered a third way: a formula for self-determination that would be implemented by means of a swift and non-communist process. The plan envisaged a programme for the training of pro-Western African elites, the constitution of political parties and a referendum on self-determination. The proposed schedule, given the urgency of the international conjuncture, should be fast-paced: political parties in 1965; referendum in 1967; and independence in 1970. Once again, compensation would be essentially economic, including significant assistance to Portuguese economy that should enable it to double its *per capita* income within five years. In case of a refusal by the Lisbon government, the plan envisaged a radical measure: Salazar's overthrow.[9] The plan was approved in March of 1962, albeit without the radical solution of the regime's overthrow and stipulating the amount of what would be the initial economic aid: seventy million dollars.

With this package aimed at solving the African question, the Kennedy administration thus launched its diplomatic offensive directed at the Lisbon government. At the same time, though, there was another pressing matter: that of the Lajes Base, whose 1957 supplementary agreement would expire by the end of 1962. In the context of the ongoing Cold War, and in the strategic framework of the so-called rapid reinforcement of Europe, the Azorean base, and hence the renewal of its facilities, was of paramount importance for the United States. The simultaneity of the two questions enabled Portuguese diplomacy to work out a strategy to connect them and, in the face of the pressure to decolonise from the Kennedy administration, Salazar responded with resistance and, later on, with a refusal to renew the agreement. Dean Rusk, the US Secretary of State, had already alerted the Portuguese Minister of Foreign Affairs to the urgency of the problem, which could lead to grave consequences in terms of its operational use. But when he came to Lisbon, in July of 1962, not only Salazar proved adamant in his position, he also directed profuse criticism at the North American positions. And even when, shortly after, Franco Nogueira visited the White House, Kennedy did not achieve better results in the face of Portuguese obduracy. The colonial question was stuck in a deadlock and the question of the Azores did not fare much better. Following Dean Rusk's diplomatic pressure in the framework of NATO, and given the urgency of the problem, the Portuguese government found a *sui generis* diplomatic solution that nevertheless did not conceal its dissent, not to say hostility, towards the great American power. Unable to drive away the US from the Azores, Salazar authorised the stationing and the use of the Azores base, but, it must be stressed, "only *de facto* and during negotiations"[10] That is, stressing diplomatically the *de facto* and not *de jure* character of the situation. This precariousness indeed extended beyond 1964, since, despite signs of moderation in the evolution of the US African policy after Kennedy, bilateral relations between Washington and Lisbon did not allow for a resumption of negotiations until Salazar's exit.

A second version of the Sakwa Plan, now drafted by Charles Bowles, was attempted in 1963. More compromising, it was presented to Salazar by the US Under Secretary of State himself, George Ball, while visiting Lisbon. He went so far as to offer 500 million dollars as financial compensation, but unflinching and terse, Salazar gave the famous reply: "Portugal is not for sale".[11]

Between the failed coup of 1961 and the attempted agreements, the US administration even asked a number of countries with whom Portugal had friendly relations – England, Spain, Brazil and the Vatican itself – to help pressure Lisbon to change its position. Portuguese foreign policy firmly resisted all their endeavours. As had occurred with the Indian question, England turned its back on the problem of African decolonisation. Spain distanced itself, on several occasions, from Portugal's positions and the same did Brazil, under the presidency of Jânio Quadros. Even the Vatican abstained from supporting Portugal, and later showed its support to the liberation movements in the Portuguese colonies.

The last US attempt at a negotiated solution for the Portuguese colonial conflicts takes place in 1965 and became known as the Anderson Plan. Designed and negotiated by the North American ambassador to Lisbon at the time, Admiral Anderson, the new plan was not substantially different, either in its goals or in its methods, from the previous ones. However, the political context and the diplomatic effort involved awarded it a much more compromising tone, and the good relations between the ambassador and Salazar himself suggested that this plan might be more successful than its predecessors. It envisaged a transition period of eight to ten years, by the end of which a referendum was to take place under international supervision, with three possible outcomes: continuation of the status quo; a Lusophone Commonwealth; or total independence. Besides the financial aid, the plan stipulated that Portugal would be granted military aid in case of non-compliance with the agreement and/or non-acceptance of the referendum results on the part of the liberation movements. Once approved by Ball and Harriman, the plan was presented to Franco Nogueira, who agreed to discuss it with Salazar. This was September of 1965. Shortly after, Salazar outright rejected it, as it had done the previous plans.

The change in the US administration, with Lyndon Johnson's rise to power and, above all, the exacerbation in the Vietnam conflict, changed, in their turn, the hierarchy of priorities in US foreign policy. In its new agenda, Africa, the colonial question and the colonies' right to self-determination had become secondary.

Despite the political nuisance and the absence of overt diplomatic support, relations – both bilateral and multilateral – with European countries, allies in the Western Bloc, were less strained and marked by a greater tolerance, and sometimes even cooperation, as was the case with France and West Germany.[12] Multilaterally, the main stage for political contestation of Portuguese foreign policy was undoubtedly the UN, its General Assembly in particular, then ruled by the spirit of Afro-Asianism and non-alignment of the new independent states freshly arrived to the international scene.

Article 73 of the Charter of the United Nations compelled member-states to transmit information concerning territories under its administration lacking self-government, that is, colonial territories. Portugal declared at once that it possessed no such territories. And, as early as 1957, a motion was proposed to condemn Portugal for not providing the required information. Portugal protested that, as a matter of political importance, it required a two-thirds majority vote. By a simple majority vote, the matter was deemed politically important, but the following vote does manage to gather the two thirds. A long period then follows of political conflict and juridical debate at the United Nations, through which Portugal's position remained constant and unyielding. Formally, according to the constitutional revision of 1951, Portugal did not possess any colonies. Instead, it was politically and administratively divided into provinces, some of which were geographically located outside Europe. It was "Portugal from Minho to Timor". In this context, any interference on the UN's part in Portuguese overseas provinces would be regarded as encroachment in Portugal's internal affairs and, thus, a violation of national sovereignty.[13] The juridical line of argument, together with the ideological discourse that highlighted Portugal's "civilising mission" and the accusation of communism directed at African nationalisms, which was not devoid of impact in the international context of the Cold War, summed up the Portuguese position.[14]

Between 1957 and 1958, Portugal invariably managed to gather that one third of the votes which precluded the motion of condemnation. However, from 1959, with the increase in the number of independent states at the UN – all of them Afro-Asian – Portugal lost the blocking minority. And, in December of 1960, two resolutions were passed at the General Assembly of the UN, which had a great impact on Portuguese colonial and foreign policies. The first, Resolution 1514 (XV) politically condemned and deemed colonialism illegal; the second, Resolution 1542 (XV), regarded the territories under Portuguese administration as colonies and urged Portugal to provide the required information. Lisbon accepted neither and the controversy escalated further.

In the framework of the contestation of Portuguese colonial policy, the positions of the Eastern Bloc and the Third World converged. Among the Western Bloc countries, however, positions varied and evolved throughout the sixties and until the end of the regime and Portuguese decolonisation. When, in December of 1960, Resolution 1542 (XV) was passed at the General Assembly, the European countries and the United States themselves opted for abstention. During the Kennedy administration, the North American position evolved into open and explicit condemnation and, in 1961, regarding the question of Angola, the United States voted for the first time, at the Security Council, on the side of the Soviet Union and against Portugal. In the midst of the Cold War, the US voted alongside Soviet Union against one of his allies within NATO. In these first confrontations, the European powers, namely Great Britain and France, maintained a moderate position. They neither supported Portugal nor the attacks directed at it, and continued to abstain from voting. During this phase, only Spain and Brazil appeared to be more sympathetic with the Portuguese position. From 1962, contestation rose in tone among the Afro-Asian countries and, at their request, formed three specialised committees to handle exclusively the question of the Portuguese colonies. From 1963–1964, the more moderate African policy of the post-Kennedy administration also had an impact on NATO, and after the overt criticism and condemnation of Portuguese colonial policy, Washington now followed its European allies in their strategy of abstention regarding the Portuguese colonial problem. Abstention that gave way to complacency with the arrival of the Nixon administration, towards the end of the sixties. At the General Assembly of the UN, however, contestation became more radical in the course of the seventies and persisted until the end of the regime. The establishment of the Organisation of African Unity (OAU), ensuing from the Addis Ababa Summit, in May of 1963, launched a new multilateral forum for contestation of Portuguese colonial policy.

The Portuguese African question also reflected in NATO and Portuguese foreign policy even had to face a number of its allies. From 1961, the military answer to the outbreak of war in Angola and subsequently in the other theatres entailed a profound change in Portugal's defence policy. The war effort and the deployment of contingents and military equipment to the African theatres of war had as a consequence the complete deferment of military commitments with NATO. The resources that were supposed to be channelled to Atlantic defence commitments were now allocated to the African theatre. Despite the above and the political discomfort that the Portuguese membership caused in some its partners, NATO represented, for Portugal's foreign policy, a diplomatic instrument that assuaged its international isolation.[15]

In May of 1961, NATO as an organisation did not raise objections, as it had done in the French case with Algeria, but the Kennedy administration, in its condemnation of Salazar's colonial policy, not only ceased all military aid to Portugal, but also demanded that NATO's war material was returned to the metropole and no longer used in the African war. The US position was further enhanced by the decree of an "arms embargo" on Portugal, which would nonetheless have more symbolic weight and political impact than practical consequences. Above all, it was short-lived. Although non-officially, US military aid was gradually resumed, and between 1963 and 1968 amounted to around 33 million dollars.

Among the European allies, France was the country that offered more military cooperation to Portugal during the war. Although France was a democracy and its international standing incomparable, the Portuguese vicissitudes echoed its own experiences of Indochina and Algeria, eliciting in it some kind of veiled sympathy. It sold equipment and armament to Portugal, namely ships, helicopters and armoured vehicles, and imposed little or no restrictions on the use of that material in the pursuance of war in Africa.[16] Bonn was another European ally that supplied arms to Portugal. As compensation, Portugal grants West Germany the use of the Beja Air Base as a training facility for the German Air Force.[17] Although closer to the US positions and, therefore, politically more aloof and diplomatically more discreet, even England never stopped supplying arms to Portugal. From 1965, however, with Rhodesia's unilateral declaration of independence and Lisbon's indirect but genuine support to the white regime, giving the interests at stake in Mozambique, Portuguese–British relations go through a period of greater tension.[18]

The turning point in Portuguese colonial policy might have arrived in the summer of 1968, with Salazar's retirement and Marcello Caetano's rise to power. Pure illusion. When Caetano, who at the outbreak of the conflicts had been one of the few die-hards of the regime to entertain the hypothesis of a political evolution towards a moderate federalism in the colonies, arrived at the head of government, he found neither the will nor the necessary political conditions to change the situation. And, despite the alternatives that sprouted in the debate between Europeanists and Africanists,[19] everything remained basically the same. The political and diplomatic options and, consequently, the African priority and the war effort.

The international conjuncture itself favoured that situation. In fact, Nixon's election, in late 1968, and, with it, Henry Kissinger's appointment as National Security Adviser, entailed a significant change in US African policy. The Cold War was starting to spread to the African continent, with the progressive influence of communism over a number of liberation movements and new independent states. In this framework, and according to the reports of the UN's Security Council, the white regimes of Southern Africa had come to stay and the strategic importance of the region made it compulsory for the US to develop cooperation in partnership with them. And, consequently, with Lisbon's government. In his self-confessed realism, Kissinger considered that, notwithstanding the authoritarian nature of the regime, Portugal was still an ally in the context of NATO and an advocate for the West on its African flank.[20]

In this sphere, one of the first measures of the new administration was to end all contact with the liberation movements in the Portuguese colonies. And, next, to partially lift the arms embargo on Portugal, which formally was still in force. This was a major change.

The new African policy of the Nixon–Kissinger duo and their concrete measures concerning the Portuguese colonial question generated a fresh atmosphere in bilateral relations. Marcello Caetano, in his turn, also revised Portuguese foreign policy with respect to the US. He strove to avoid ideological confrontation and political tension, and to resume normal relations and a degree of cooperation with the US administration. In the context of this new atmosphere, as early as 1969, the Portuguese government conveyed to Washington its interest in solving the still pending question of the Lajes Base in the Azores and proposed the opening of negotiations. Washington, on the other hand, in 1970, following this demonstration of goodwill, lifted the embargo, with immediate benefits for the Portuguese Armed Forces and the war effort in Africa. Supply of weapons, ammunition, defoliant chemicals and even specialised training in counter-guerrilla techniques for the Portuguese military were authorised by the USA.

The Lajes Agreement was finally signed following a NATO meeting in Brussels, in December of 1971, and was in force until the transition to democracy, in 1974.[21]

Until 1974, however, one other international event of the utmost importance had a positive impact on Portuguese–American bilateral relations, with immediate effects on the decolonisation conflicts: the Yom Kippur War between the State of Israel and the Arab countries, in 1973. In order to deliver emergency support to Israel, the US Armed Forces required a strategic refuelling point for their airlifters, a facility that no other European country showed availability to grant. In this context of crisis and faced with the strategic imperative of the US, Lisbon sets in motion an "opportunity-based diplomacy": it made the use of the Azores Base conditional on the supply of a list of weapons, especially surface-to-air missiles, that the Portuguese Armed Forces were in desperate need of for the conduct of war in Guinea. It proved ineffective. The US rejected the political proviso of the Portuguese proposal and, under pressure, Portugal was forced to grant the requested authorisation. Once again, the Azores will prove strategically crucial for the US forces.[22] Despite the formal rejection of the Portuguese claims, Lisbon's attitude had a positive effect on bilateral relations. So much so that Kissinger endeavoured to provide the supply of the much-craved missiles, which would obviously reach Portugal through a "devious way"[23] and whose delivery, through West Germany, was scheduled for April of 1974. Too late, as it turned out.[24]

Despite the Western Bloc's *realpolitik* and the fact that they had left the top of the international agenda, the colonial question and the decolonisation conflicts in the Portuguese colonial territories had suffered no change in the first half of the seventies.

At the bilateral diplomatic level, the Portuguese position was dealt a huge blow in July of 1971. After his attendance, in 1969, at the Kampala Episcopal Conference, which the Portuguese clergy were barred from entering, Pope Paul VI decided to grant an audience to the representatives of the liberation movements of Portuguese colonies: Agostinho Neto, Amílcar Cabral and Marcelino dos Santos. The Vatican's attitude had profound political and moral repercussions in Portugal, given the Catholic values cherished by the regime and the influence of the Church in a mostly Catholic population.

Multilaterally, there was a progressive exacerbation of positions, invariably encouraged by the Afro-Asian group. In 1971, Portugal left the UNESCO. And, at the UN, Portugal's position was increasingly more difficult to uphold. After a visit to Guinea's "liberated zone", the UN's Special Committee on Decolonisation recognised the PAIGC as the sole representative of the people of Guinea-Bissau, a resolution adopted by the General Assembly on its 27th session, in October of 1972. Less than a year later, in September of 1973, in Medina do Boé, the PAIGC declared unilaterally the independence of Guinea-Bissau. In a little over a month, eighty UN countries had officially recognised the new State. In October, the 28th session of the General Assembly of the UN formally condemned Portugal's "illegal occupation" of Guinea-Bissau.

Around the same time, international public opinion was yet again faced with the Portuguese colonial question and the military conflicts of decolonisation. Not for the first time, or even the second, it fell under the spotlight of the media. On this occasion, however, it would deserve its greatest media coverage. In the London *Times*, an English Anglican minister, Adrian Hastings, published an article denouncing a massacre of civilians perpetrated by Portuguese troops in Mozambique: the Wiriamu massacre.

In this phase, however, the core of the conflict could no longer be decided in the ideological debate, through diplomatic support or even in the arena the international public opinion, but rather on the ground, in military terms.

Mobilising for war: Defence policy, the armed forces and guerrilla warfare

After political and diplomatic confrontation, the time had come, for Portugal, for military conflict. The war played out in three different African theatres: Angola, Guinea and Mozambique.

Defence policy entered a new and completely different phase. The shadow of the continental threat, prevalent theretofore, had been left behind. Thenceforth, the new threat was located in the African theatre and took on a non-conventional shape: guerrilla warfare.

For thirteen long years, this was the picture of the decolonisation conflicts: a non-conventional war, a guerrilla war, which, between 1961 and 1974, monopolised all of defence policy. Africa became top priority and this priority was directed towards counter-guerrilla warfare, according to the model of the French experience in Algeria.

Since 1959, the first signs of change in defence policy had begun to take shape. In a document drafted by Defence Minister Botelho Moniz and approved by the Supreme Council of National Defence in the same year, compliance with NATO military commitments was still upheld.

But only theoretically, because in practice the focus of priority had already begun to shift towards Africa.

Ten years of NATO membership, and under its training model, had bestowed on the Portuguese military elite autonomous capacity in terms of information and strategic analysis. Therefore, between 1959 and 1961, that generation of NATO military elite, led by the Defence Minister, General Botelho Moniz, developed a strategic study that clearly advised against pursuing a military option to solve the colonial question. In April of 1961, Botelho Moniz's attempted coup had precisely that strategic objective: to remove Salazar from power, to change colonial policy and to opt for a non-military solution, that is, a political solution for the decolonisation conflicts.

When the coup failed, any chance for a negotiated solution for the colonial question also vanished. Salazar feared that yielding with respect to the autonomy of the colonial territories, even if from afar and indirectly, might have a political impact on the internal evolution of the regime itself. He was not wrong either. The persistence of the empire depended on the continuity of the regime, and the persistence of the regime depended on the continuity of the empire. From that time onwards, Portugal's chief goal was the defence of the colonial *status quo*. The new military hierarchy, ensued from the 1961 *Abrilada*, supported this goal at the political level and carried it out at the strategic and military levels. Almost all commitments with NATO were suspended and the defence policy directed all its efforts to counter-guerrilla warfare in the African theatre. "To Angola, swiftly and powerfully!", said Salazar.

The war effort in Africa entailed important changes in the military institution, at various levels: conscription and personnel, organic structure, armament and, naturally, military doctrine.

In the first place, the issue of recruitment. The war effort called for a tremendous increase in active duty personnel, requiring the recruitment and mass mobilisation of militia soldiers. For the first time since the establishment of the conscription system during the First Republic, military service in Portugal became truly universal and compulsory. In thirteen years of war, the total number of active duty personnel in the Portuguese Armed Forces increased by five and a half times. In 1961, the total number of soldiers neared 40,000 men, whereas in 1974 it stood at 217,000. By the end of the conflict, in 1974, the total number of mobilised soldiers in the three operational theatres was around 149,090. In the course of the thirteen years that the war lasted, the average number of soldiers in the three theatres is 117,000, of which 107,816 in the Army alone.[25] The recruitment effort encompassed not only private soldiers, but also officers and sergeants. In the face of the inexhaustible need for human resources that the war entailed, a variety of mechanisms were employed to address it.

First, an increase in the length of service: six months of training followed by a two-year deployment to the operational theatres, which was sometimes further extended.

Second, stricter recruitment procedures, which practically put an end to any kind of exemption. Only resorting to illegal means was it possible to evade service. In 1961, the percentage of men drafted for service represented 64.8% of the total number of the registered, and, in 1972, it stood at 72%. Nevertheless, the number of draft evaders kept rising throughout the whole duration of the wars. It stood at 11.6% of the total number of men drafted for service in 1961 and grew gradually and steadily to 20.3% in 1974.[26]

Third, the Africanisation of troops, that is, the extension of recruitment to African populations, from the second half of the sixties, but especially from the seventies. When the war began, in 1961, the average percentage of African troops in the framework of Portuguese troops stood at 20.9%; by the end of the war, it had all but doubled, standing at 38.7%. The numbers, however, varied greatly according to the theatre in question: whereas the percentage almost triplicates in Angola, from 149% to 42.4%, in Mozambique it doubles from 26.8% to 53.6%, and in Guinea it remains almost unchanged, with a slight decrease from 21.1% to 20.1%.[27]

Fourth, successive measures respecting the cadre of officers: the call-up of militia soldiers that had already rendered service in the metropole and had transitioned to the reserve; the creation of a special cadre of officers, with a selection of cadets from the courses for militia officers to enrol in intensive training and be given swift promotion to the rank of captain; and, finally, legislative changes that altered the seniority criteria of officers in the permanent cadre, between those coming from the Military Academy and the ones from the militias. This last measure would in fact be at the root of the discontent, between the captains, that would ultimately give rise to the Armed Forces Movement.[28]

The recruitment and mass deployment to the colonial theatres would have huge repercussions not only in the Armed Forces, but also in Portuguese society as a whole and particularly in Portuguese youth. Between 1961 and 1974, around one million Portuguese young men – one percent of the Portuguese population – did their mandatory military service at the colonies. Comparing its impact on society, suffice to say that, on average, the United States committed only 0.25 percent of its population to the Vietnam War, whereas in Portugal that number stood at one percent.[29]

In the second place, the issue of organic structures. The long duration of the war and its multiplicity of theatres required the military institution to adapt its organic structure. Keeping up the war effort in three different theatres, distant from one another and all of them distant from the metropole, entailed from the outset a territorial reorganisation of the Army. In 1962, new military regions are created: Angola, with five territorial commands, and Mozambique, with three territorial commands. Also in 1962, the territories of Guinea, Cape Verde, São Tomé e Príncipe, Macau and Timor became autonomous territorial commands.[30] The purpose was clear: to enable a more flexible command and greater speediness and effectiveness in the conduct of war. On the other hand, the need to adapt the military apparatus to counter-guerrilla warfare led to the implementation, in the areas of operational activity of the various theatres, of a square division whose base unit was the battalion and which depended militarily on a chain of command whose hierarchy ranged from the sector command to the zone command, the regional command and finally the chief-command. Under specific circumstances and as a requirement of certain missions, further operational commands had to be created. Although during the war there was no global reform of the Army to speak of, a number of measures were nonetheless taken to meet war requirements in a swifter and more efficient way. To begin with, the creation of Transmissions and a number of new units – the Service Battalion, the Transmissions Reconnaissance Battalion, the Medical Regiment and the Military Postal Service. But more important still, and with a greater impact on the conduct of war, was the formation of troops especially intended for guerrilla warfare. In 1960, the Special Operations Centre was established in Lamego and, from 1962, the training and employment of special troops – the commandos – began in Angola, which were used later on also in Guinea and Mozambique. Paratroopers[31] and *fuzileiros* (Portuguese Marines) will also make their debut in the three operational theatres. With the growing Africanisation of the Portuguese Armed Forces, from 1965, in the various theatres, a number of special units were created which were exclusively composed of African troops: the Special Groups (GE) and the Special Paratrooper Groups (GEP), the African Commandos, the Special Militias and the Arrows, the latter being the more controversial force, due to its composition, methods and articulation with the PIDE.[32]

The nature of the conflicts – guerrilla and counter-guerrilla warfare – conferred to the Army a central role in Portuguese military manoeuvring. However, the Navy and the Air Force also played an important part in providing logistic support and strategic and tactical mobility, which required these two branches to carry out, on their part, a number of functional and organisational adjustments.

The Armada also underwent come organic restructuring in order to comply with the new challenges posed by the decolonisation conflicts.[33] In 1957/1958, two new naval commands are created: one in Angola and one in Mozambique. The new naval and maritime defence commands, which gathered, within the Navy, military and maritime authority functions, were endowed with autonomy precisely to be able to respond more effectively and promptly to the new missions faced by the Armada:[34] to ensure logistical communications of a strategic nature between the metropole and the theatres of war; to control the coastlines and tactical and logistical support lines in the littoral, as well and lines of fluvial penetration in the operational theatres; containing infiltration in those areas; and carry out amphibious operations. In 1961 they were incorporated in the Navy ranks and the Marine Corps became fully operational.

The Armada went through substantial growth, during this period. Whereas in 1960 it had a force composed of 71 ships with a tonnage of 64,332 t, in 1973, the number of ships had risen to 186 with a tonnage of 84,732 t. That represents an increase of roughly 33% in 13 years. Obviously the great majority of these ships were patrol boats and landing craft, suitable for the African theatres.[35]

The Air Force also registered, around the same time and to the same strategic purpose, a structural readjustment.[36] As had been the case with the Armada and the new naval commands, since 1956 there were plans for the creation of two new Air regions, in Angola and Mozambique, with a central air base in each one them, Luanda and Lourenço Marques, which would to be supported by a network of auxiliary bases. When war broke out, however, this structure was far from operational. The chief missions of the Air Force were also well defined, being primarily to provide strategic transport between the mainland and the African theatres and tactical transport and mobility in the operational theatres. The fleet assembled during the fifties, in the framework of NATO and intended for Atlantic defence, was not adequate to this end. The Air Force thus had to undergo a renewal of equipment and armament.

The third place is precisely that of equipment and armament. Whereas during the fifties, in the framework of NATO and under the US-Portuguese agreement on Mutual Defence Assistance, the renewal of the Portuguese Armed Forces was carried out under the influence of the United States, the sixties brought a radical change to the panorama. As a result of the position of the Kennedy administration regarding Portuguese colonial policy and the embargo on arms then decreed, Portugal was forced to seek new solutions. New alliances were then essayed in the field of defence, and new suppliers were sought for the rearmament of the Portuguese Armed Forces: France, Germany and, to a lesser extent, Spain. Mainly, though, there was tremendous development in national military industry, which largely sustained the war effort during the decolonisation conflicts, namely in terms of equipment, light weapons and ammunition.[37] With regard to the Army, for instance, a variety of weapons were manufactured in Portugal: G3 assault rifles (7.62 mm), a NATO weapon developed by the German, HK21 machine guns (7.2 mm), FPB submachine guns (9 mm), 60 and 81 mm mortars and machine-gun Chaimite armoured vehicles. Portugal also manufactured *dilagramas* m/65 and hand-grenades. The National Factory of Light Weapons Ammunition and the Braço de Prata Military Factory were the Portuguese military industry establishments that underwent a greater development. Not only did they meet the requirements of the Portuguese war effort, they also supplied other NATO allies. With respect to logistics, equipment, uniforms, rations and health care, establishments such as the Uniforms and Equipment General Workshops, the National Rope Factory, the Military Maintenance and the Chemical and Pharmaceutical Military Laboratories underwent significant development.[38] The shifting alliances in the field of defence, with the resulting changes in weapon suppliers, also affected the Armada. The new ships for the Navy were no longer purchased from the United States or Great Britain, but rather from France, Germany and Spain. Moreover, just like with the Army, the Armada also registered a huge development in national manufacture. According to a project and under the technical supervision of the Naval Construction Inspection, a large number of ships are built by Portuguese naval industry, at the Lisnave, Mondego or Viana de Castelo shipyards. It was the case of the patrol boats and landing craft used in Africa, such as the Argos, Jupiter and Cacine. The new submarines and frigates, namely the João Belo type, were of French origin, while the corvettes and other types of speedboats, such as the Bellatrix, were of German origin.[39]

The re-equipment of the Air Force followed similar lines. According to an assessment carried out at the time, the most effective technical solutions for the missions of strategic and tactical air transport that the Air Force would have to face in the African theatres, in the scope of the decolonisation conflicts, were of American origin: the DC-6 and the DC-7, the C-130 Hercules and the Bell UH-1 Iroquois helicopter. However, aware that the aircraft was intended for the African theatres, as early as 1958 the US refuse to cooperate, a refusal which was subsequently iterated and reinforces by the embargo decreed by the

Kennedy administration. Portugal and the Air Force then turned towards European alternatives in France and Germany and Portugal received its order of French Alouette II helicopters and Noratlas aircrafts, Broussard monoplanes, German DO-27 aircrafts, and, from 1963, a reinforcement of the Noratlas fleet and Alouette III helicopters.[40]

During the war years, military expenditure soars, reaching the highest numbers of the entire 20th century, on a par only with those registered during the Second World War. Prior to the onset of the conflicts, in 1959, Portuguese military expenditure amounted to 4% of the GDP. When the conflicts peaked, in 1971, the number had doubled to 8% of the GDP, proportionally much more than what the US had spent with Vietnam.[41] One other significant index is the war budget. At the outset of the war, in 1961, military expenditure stood at 36% of the total public expenditure. Between 1965 and 1967, war spending already neared the 40%, peaking in 1968 at 42.4%. In 1973, by the end of the war, it still stood at 32.6%. It is no easy task to ascertain what share was allocated to each branch of the Armed Forces, insofar as the totality of the expenditure falls, as pertaining to the war effort, under the budget item "national defence general expenses". It is rather obvious, however, given the relative weight of the forces in the ground, that the largest share was allocated to the Army.

Lastly, there is still one domain regarding which the decolonisation conflicts imposed on the Armed Forces the opening of a new cycle: that of doctrine, training and guerrilla and counter-guerrilla warfare techniques. From 1957/1958, Portuguese officers begin to enrol in counter-guerrilla courses with the Belgium armed forces. It was under France's influence and experience in Algeria, though, that the Portuguese Armed Forces began their training – regarding both doctrine and cadre practice – at Algeria's Guerrilla Warfare School. Later on, under the Nixon administration, the United States agreed, reluctantly and restrictedly, to offer military cooperation in this field, allowing the attendance of Portuguese officers at their counter-guerrilla warfare schools in the Panama Canal Zone. Even more important, though, was the fact that the military experience afforded by thirteen years of war in three different African theatres had prompted a reflection that outdid its French and American sources, producing an original strategic thought and a distinctly Portuguese doctrine in terms of guerrilla and counter-guerrilla warfare, or "subversive warfare", to use the official expression. That strategic doctrine was devised at the Institute for Higher Military Studies in Pedrouços (Lisbon), and its most accomplished instance, the regulation titled "The Army in subversive warfare" was used, not only by the Portuguese Armed Forces, but also by the US ones in the Vietnam War.[42]

Independence movements and armed conflict: One or three decolonisation wars?

Colonisation and decolonisation[43] are two sides of the same coin. As such, Portuguese decolonisation begins with the revolts against colonisation itself. However, in line with the modern definition of the concept, resistance against colonisation and the formation of anti-colonial movements at the Portuguese colonies began with the shaping of African elites in the second half of the forties and the emergence of independence movements in the fifties and sixties.

The elites were mostly recruited among the small segment of "culturally assimilated" populations in Angola and Mozambique, and, in Guinea's case, under a strong Cape-Verdean influence. A part of these African – often mixed-race – elites had studied at and graduated from Portuguese universities, and even participated in unitary movements against the *Estado Novo*, as was the case of the MUD-Juvenil, in the second post-war period. The Home for the Students of the Empire,[44] connected to the Portuguese Youth, and whose function was precisely to house and support those African university students in the metropole, was paradoxically the institution that most contributed to hosting and divulging anti-colonial culture in Portugal. Throughout the forties and the fifties, many of the leaders of the future liberation movements and independent states stayed there: Amílcar Cabral and Vasco Cabral, from Cape Verde, Agostinho Neto, Lúcio Lara and Mário Pinto de Andrade, from Angola, and Marcelino dos Santos, from Mozambique. Others followed alternative paths, from the end of the fifties. They sojourned briefly in Portugal and then left, legally or illegally, to other European countries or the United States, where they pursued their studies and developed their activity. It was the case of Eduardo Mondlane, from Mozambique, or Jonas Savimbi, from Angola.[45]

With the development and worldwide affirmation of the Afro-Asian movement from the mid-fifties, these African students who, in Portugal, had militated with the Portuguese opposition in unitary movements against the Portuguese dictatorship, begin to organise themselves autonomously and to claim the representation of their homelands at international meetings. In 1953, in Bucharest, in a student congress for peace, Agostinho Neto introduced himself for the first time as Angola's representative, while Marcelino dos Santos did the same for Mozambique and Vasco Cabral for Guinea and Cape Verde.[46] The cycle of the Portuguese colonies' independence movements had begun.[47]

In Guinea, in September of 1956, led by Amílcar Cabral and heir to the former Movement for the National Independence of Guinea (MING), the African Party for the Independence (PAI) was born. Subsequently, in exile, it would take on the designation of African Party for the Independence of Guinea and Cape Verde (PAIGC), invariably supported by its Francophone neighbours of Guinea-Conakry and Senegal. In 1958, the National Union of Workers of Guinea-Bissau was founded, a budding and clandestine union movement.

In Angola, the process was different and the model of independent movements had a two-fold origin that reflected its internal diversity and rivalries: on the one hand, the movements based on the mobilisation of urbanised and mixed-race African populations settled in Luanda and led by cosmopolitan elites forged by Portuguese universities; on the other, the ethno-nationalist movements, based on specific ethnic groups and limited to their territorial area.

Among the former were the Angolan Communist Party (PCA) – which started as a cell of the Portuguese Communist Party (PCP) in 1948 and was founded as a proper party in 1955, with Viriato da Cruz serving as the connecting link with its Party of the United Struggle for Africans in Angola (PLUAA) founded in 1956. As a result of the merging between the manifesto of this group and the Movement for the Independence of Angola (MIA), it came out, in 1956, that which is considered to be the founding proclamation of the People's Movement for the Liberation of Angola (MPLA), led by Viriato da Cruz and Mário Pinto de Andrade. This designation, however, appeared at a later stage and the party, as an organisation, was established only in 1960, when exiled in Conakry.

Among the ethno-nationalist movements, the first and more influential one was the Union of the Peoples of Northern Angola (UPNA), almost exclusively composed by the *baconga* ethnic group and founded in 1955 by Barros Nekaka. Its politically broader ambitions led to its shedding of the "Northern" part of its name, becoming the Union of the Peoples of Angola (UPA). Nekaka's nephew, Holden Roberto inherited, as tradition prescribed, through matrilineal descent, the leadership of the movement which, in 1962, took on the designation of National Liberation Front of Angola (FNLA). Supported by neighbouring ex-Belgian Congo, it sets up, a while later, the Revolutionary Government of Angola in Exile (GRAE). One other relevant ethno-nationalist movement was the Movement for the

Liberation of the Enclave of Cabinda (MLEC), founded in 1960 by Luís Ranque Franque. In 1963, this movement merged with the Action Committee of the Cabinda National Union (CAUNC), forming the Front for the Liberation of the Enclave of Cabinda (FLEC).

Even more important and consequential was the foundation of still one other ethno-nationalist movement. Later on, in the midst of war, Jonas Savimbi, who had been Foreign Affairs Minister of the GRAE, abandoned his office in 1964 and, with the *ovimbundu* ethnic group as its basis and backed by newly-independent Zambia, launched the National Union for the Total Independence of Angola (UNITA), in March of 1966.

In Mozambique, there were also a number of ethnic-based movements, but their formation was largely encouraged and supported by the newly-independent bordering powers. In 1959, it was founded in Tanzania the ethnic-based – essentially *makonde* – Mozambique African National Union (MANU). In 1960, the National Democratic Movement of Mozambique (UDENAMO) was founded in Rhodesia, led by Marcelino dos Santos, and later transferred to Tanzania. Finally, in 1961, the National African Union of Independent Mozambique was founded in Malawi (UNAMI). These are the three movements that, under strong international pressure, particularly from new African leaders such as Nyerere and Nkrumá, were merged in Dar es-Salaam, in June of 1962, giving rise to the Mozambique Liberation Front (FRELIMO), led by Eduardo Mondlane.

In 1961, the Committee for the Liberation of São Tomé e Príncipe (CLSTP) was founded.

Separated by geography and history, these liberation movements had at least one thing in common: the colonising power. Without relinquishing their distinctive features, they endeavoured to find ways to coordinate their anti-colonial political and diplomatic initiatives. The international situation, marked by the successive waves of African decolonisation processes, the reference to the same colonial power and the sociability between some of those pro-independence elites favoured those attempts at political cooperation and diplomatic coordination. One of the first organisations embodying this cooperation was the Anti-Colonial Movement (MAC) founded in Lisbon and Paris by Agostinho Neto, Lúcio Lara and Viriato da Cruz representing Angola, Marcelino dos Santos representing Mozambique, Amílcar Cabral representing Guinea and Cape Verde, and Guilherme Espírito Santo representing São Tomé e Príncipe. From this first movement, in 1960, at the All African Peoples' Conference in Tunis, the African Revolutionary Front for the National Independence of the Portuguese Colonies (FRAIN) was created, encompassing not only the personalities, but the organisations themselves, in the cases wherein the liberation movements already had an institutional existence: Angola and Guinea. In 1961, the Conference of Nationalist Organisations of the Portuguese Colonies (CONCP) was set up, grouping the liberation movements of all the colonies (MPLA, PAIGC, UDENAMO and, later on, FRELIMO, CLSTP), which then held its first conference and based its headquarters in Rabat.

All these movements, in their declarations of principles and in the political statements of the various conferences, expressed from the outset the purpose of carrying out a "peaceful dissolution" of Portuguese colonialism, that is, the search for a political and diplomatic solution for decolonisation. Petitions and open letters to the Portuguese government, in September of 1960 and October of 1961, bumped time and again against a wall of silence. Lisbon's colonial policy, its obdurate defence of the empire and its repression of the initially peaceful pro-independence movements would thus lead to radicalisation and to an alternative course: that of armed struggle and guerrilla warfare.

The armed struggle conducted by the liberation movements, however, was preceded by a stage of more or less spontaneous popular insurgency against those which were deemed the social grievances imposed by the colonial status, and which, while appearing to be old-fashioned uprisings, proved to be modern revolts that contained in themselves the seed of independentism.

In 1953, in São Tomé e Príncipe, a first big upheaval took place, labour-related but with racial conflictual nuances, between the island's natives and the workers from Angola and Mozambique. The "monkey hunt", as it became known at the time the violent suppression of the riot, resulted in around one thousand dead in a little over one week.

In 1956, a strike movement among farm workers announced itself in northern Angola. In Mozambique, in the same year, the dockers at the Lourenço Marques Port also went on strike. This strike too was brutally crushed, resulting in forty nine deaths.

In 1959, in Guinea, labour and pay rise claims voiced by the striking dockers at the Pidjiguiti docks degenerated into mutiny. Policy suppression resulted in between seven and fifty dead and between fifteen and one hundred wounded, according to different estimates.

In 1960, in northern Mozambique, an uprising against compulsory cotton production was brutally suppressed by colonial authorities, resulting in the death of around five hundred Africans.

Finally, in early 1961, in Lower Cassanje, Angola, a new revolt against compulsory cotton production resulted in around one hundred deaths.

Portugal's adamant refusal, at the political level, to engage in any sort of dialogue or negotiated solution to the problem of decolonisation, and the violent suppression of protests and uprisings, on the ground, by Portuguese colonial authorities, eventually caused nationalist movements to turn into liberation movements and convert anti-colonial resistance into armed struggle. In three operational theatres: Angola, Guinea and Mozambique.

Angola

The decolonisation conflict in Angola, as later also decolonisation itself and the construction of the independent State, were marked by dissension among the various liberation movements at work. Perhaps this is why, in the origin myth of national liberation, the Angolans cannot agree on whether the triggering the foundational event was the Luanda prison assault, on the 4th of February of 1961, claimed by MPLA, or UPA's attacks on farms and administrative posts on the 15th of March of the same year. The former, which except for the death of seven policemen, did not affect the white population and met with resounding failure, received a great deal of attention by the international press. This was probably due to the simultaneity of Henrique Galvão's hijacking of the Santa Maria cruise liner, and the speculation, encouraged by the press, that he might be heading towards Luanda to establish a government in exile. The latter, on the contrary, was a tribal attack that resulted in a true massacre of white populations, eliciting a proportionate response by the Portuguese government. The failure of the former and the violence of the latter were the first signs that this war would not be an urban guerrilla, but rather a rural guerrilla conflict. In August of 1961, the first troops arrived in Angola. For Portugal, the first decolonisation conflict had started. In October, all the administrative posts in all towns were reoccupied. Discouraged, the UPA almost disbanded. The Commander-in-Chief General Venâncio Deslandes declared the end of all military operations. He was convinced that any loose ends might be tackled through police operations. He was mistaken. This was just the beginning and the war was there to stay.

Between 1961 and 1964, the UPA – FNLA from 1962 – appeared to be the prevalent movement, backed by the US, with a government in exile and engaged in few but persistent guerrilla actions in northern Angola. During this period, the MPLA was facing internal dissension and difficulties and scarce external support. Between 1964 and 1970, on the contrary, it was the FNLA that crossed a period of crisis marked by tribalism, Holden Roberto's autocratic leadership and internal strife, accompanied by a decrease in international support. By that time, it was the MPLA that, under the leadership of Agostinho Neto, not only began to expand its internal support basis, but also its international support. With help from Congo-Brazzaville and, from 1966, of the Organisation of African Unity (OAU) itself, and the technical cooperation of the Soviet Union and other socialist countries, the MPLA managed to become the most active movement in guerrilla actions. Between 1964 and 1966, its military activity unfolded mainly in the Cabinda enclave, but in 1966 it succeeded in opening another front to the east, led by Daniel Chipenda.[48] In 1966, the UNITA, newly-founded and led by Jonas Savimbi, also joined the armed struggle and, following a series of attacks on administrative posts in the eastern area of the territory, focussed its operational activity on the assaults on the Benguela railway.

Between 1970 and the end of the conflict, the FNLA's guerrilla actions persisted, with no great achievements, while the MPLA was once again plunged into internal strife, with grave repercussions in terms of external support and military effectiveness. The UNITA, for its part, had exhausted its already scant military capacities in fighting the other liberation movements, and ended up signing a truce with the Portuguese forces in 1972.

If there was any successful force, in military terms, it was the Portuguese Armed Forces, now under the command of General Costa Gomes. Tribalism, internal dissension within the movements, fratricide struggle in their midst and the dispersion of political leadership among the various war fronts, in addition to the military effectiveness of the "cleansing operations", all combined to make the job easier for the Portuguese forces.

Towards the end of the war, the nationalist movements posed no real military threat to the Portuguese presence in Angola, in terms of the control neither of the populations nor of the territory. The liberation movements, on the whole, claimed back control over around three million people. The numbers of the Portuguese authorities sway between 85,000 and 115,000 people under the control of those same movements, which meant 2% of registered citizens.[49] In any case, strategic interest and urban centres which were the hub of European population were protected and impervious to the conflict. In thirteen years of war, Portugal kept engaged, in Angola, an average force of 54,000 men, having registered an annual average of 105 killed in action.[50] It was, therefore, a low-intensity conflict. The Portuguese forces being in control of the military situation in Angola, the ongoing war was, as far as the white population was concerned, a far-off reality.[51]

Guinea

Much different was the situation in Guinea-Bissau, where the PAIGC was conducting a successful strategy in the fight for liberation, both at the political and diplomatic level and in military terms. A variety of geographic, social and political factors contributed to it. Firstly, the geography and ecology of the territory, with its flat, humid and swampy terrain, criss-crossed by countless waterlines, whose *bolanha* (rice fields) proved to be favourable to the conduction of guerrilla warfare. Secondly, its reduced economic value and the almost inexistent penetration of white settlers, which during the war years were never above the 2,000 people, mostly public servants rather than genuine settlers. Finally, above all, the existence of a single liberation movement endowed with a strong political leadership which managed to conduct a diplomatic action that garnered international sympathy and support and a relatively successful military strategy. Naturally, the ethnic element was present, both among the populations and within the PAIGC. It was obviously exploited by General Spínola's policy, but never in a way as to undermine the action of the liberation movement.

In guerrilla warfare, strategy unfolds in successive stages and armed action is initiated only in its last stage.[52] Between 1961 and 1963, the PAIGC undertook a series of sabotage actions in inland Guinea, but the outbreak of the armed conflict, in the classical sense, occurred on the latter date. Following a political declaration by Amílcar Cabral, in December of 1962, the PAIGC launched an attack, in January of 1963, on the Portuguese garrison at Tite. This raid symbolically heralded the beginning of the guerrilla phase and the opening of one more war front in the Portuguese decolonisation conflicts.

In Guinea, the guerrilla progressed initially from the south, alongside the border and with the support of Sekou Touré's Guinea-Conakry. The first front was opened, in early 1963, between the border and the Geba River. By the end of the year, a second one was opened, between the Geba and Cacheu Rivers, and, in 1964, two more fronts, in Gabu and south of Boé. With all these active fronts, the guerrilla penetrated into the territory, spreading inland and, by that time, the PAIGC claims control over half the territory. The Portuguese military command had recognised their control in 1963, but only over 15% of the territory.[53] In 1965, the war had reached a plateau, which did not prevent the opening of a northern front, starting from the border and with support from Leopold Senghor's Senegal.

General Arnaldo Schultz, military Commander-in-Chief between 1964 and 1968, followed a defensive military strategy based essentially on being in command of the fortified positions and gaining aerial control of the territory.

The arrival of General António de Spínola – at once Governor and Commander-in-Chief of the Armed Forces – in Guinea signalled the adoption of a radically new strategy, which would represent the biggest political-military challenge that the PAIGC's action had to face. Knowledgeable about the Algeria and the Vietnam experiences, Spínola was aware that guerrilla warfare is not fought, let alone won, in strictly military terms.[54]

A charismatic man, he embarked on a strategy of "psychological action" and invested in the economic and social development of the populations, reassembled in villages and protected by Portuguese military forces. He even went so far as to launch a structure of political representation in direct defiance of the PAIGC. In the first place, he placed Portuguese military resources at the service of the indigenous populations, in the construction of public works and the building of health and education infrastructures. On the other hand, he created formal assembly-like structures that worked as advisory bodies, the so-called People's Congresses, which, by engaging the tribal elites, challenged the para-legislative systems that the PAIGC had established in the "liberated zones". Finally, Spínola did not shy away from playing the ethnic card by supporting anti-Cape Verdean organisations and promoting ethnic groups less permeable to the guerrilla, particularly the *fulas*.[55]

At the military level, he authorised a more aggressive strategy, the bombing of PAIGC villages and bases on the other side of the Senegalese border, and even special operations such as the well-known Operation Green Sea, in December of 1970. Green-lighted by the Lisbon Government, albeit with reservations and on the explicit condition that no traces should be left – the operation consisted in an attack on Guinea-Conakry with a threefold objective: to overthrow Sekou Touré and establish, in Conakry, a regime favourable to Portugal; the destruction of Guinea-Conakry's aerial resources; and, finally, the liberation of Portuguese prisoners. It proved to be a relative military failure and a total political fiasco. The operation managed to rescue the Portuguese prisoners and to lay bare the vulnerability of the guerrilla sanctuaries. But, not only did it not succeed in destroying Guinea-Conakry's air force, it also left the inerasable mark of its identity. Moreover, it had an unwanted effect: it failed to overthrow the establishment in Conakry and led Sekou Touré to ask for the assistance of the Soviet Union in patrolling the West African coast.

In an effort to overcome the deadlock that gripped the conflict, Spínola also attempted, at the diplomatic level, to open indirect negotiations with the PAIGC, through the mediation of Leopold Senghor. Also unsuccessfully, since the initiative was promptly blocked by the government in Lisbon.

Under circumstances that have never been explained, Amílcar Cabral was assassinated in January of 1973. But not even the death of its historical leader, and his replacement by Aristides Pereira as head of the movement, hindered the PAIGC's guerrilla, which would soon register two decisive accomplishments: one at the military level and the other at the political and diplomatic level. At the first level, the Soviet Union supplied the PAIGC with SAM-7 ground-to-air missiles, whose role in the operational theatre was decisive, since they were able to neutralise the one element of military superiority of the Portuguese forces: air power. At the second level, taking advantage of the strong international support it was granted, the PAIGC unilaterally declared Guinea's independence, which was promptly recognised by 80 States at the United Nations. These States, moreover, condemned Portugal's illegal occupation of the territory and recognised the PAIGC as the sole legitimate representative of the people of Guinea-Bissau.

As far as Portugal was concerned, the situation in Guinea was critical at all levels: political, diplomatic and military. The PAIGC posed a real threat and, in the last years of the war, controlled around a third of the territory and disputed with Portugal a second third, while Portugal controlled the last third, which encompassed Bissau, Bolama and Bijagós.

During the war years, Portugal employed in Guinea an average force of 20,000 men, having registered a total of 1,342 soldiers killed in action, which meant an annual average of 122 killed in action.[56] Taking into account the low economic value of the territory and the low level of social penetration of Portuguese colonisation, this meant for the country an enormous war effort – a terribly heavy financial, political and human burden. This was not about costs versus benefits, however, but about precedents. As had happened with India, if the Estado Novo set a precedent with Guinea, what would it do with Angola and Mozambique?

And still, at the eleventh hour, faced with a very critical situation in Guinea and after having blocked Spínola's indirect negotiations with the PAIGC in 1972, Marcello Caetano himself attempts to negotiate directly with the liberation movement. In London, in the secrecy of a hotel room. It was too late, however. The first conversation takes place days before the coup of April of 1974. There would never be a second one.[57]

Mozambique

In Mozambique, the outbreak of war was not much different. After the political and diplomatic endeavours, and the initial stages of preparation and stirring of the guerrilla, the outset of the armed struggle is signalled by an attack on Mueda, in the Cabo Delgado district, in September of 1964. In the first phase of the conflict, the FRELIMO extended the guerrilla, starting from Tanzania and with the support of the *makonde* ethnic group, to the whole north of Mozambique, the Makonde Plateau. One more operational theatre and a third decolonisation war had been launched, for Portugal.

However, north to south penetration was not easy, due to the presence of the Makua people, ethnically and religiously different from the Makonde. The FRELIMO therefore sought an alternative strategy once again relying on Tanzania's support but now also Malawi's, expanding the guerrilla towards the west in the Nyassa zone.

Ambushes, mines and attacks on the barracks gradually became more frequent. From a military point of view, though, the actions of the FRELIMO cannot be described as particularly successful. On the contrary, the fact that war broke out at a later stage, in Mozambique, awarded the Portuguese forces – which had accumulated the experience of the other theatres and were thus able to respond more quickly and effectively – time to prepare and mobilise. In 1965, the intensity of FRELIMO's armed struggle had dwindled significantly in the north, and repeated attempts to expand it to the south – to Tete and Zambezia – had been botched. On the other hand, this relative military shortcoming and the defections that followed had consequences in FRELIMO's in-house affairs, which went through a period of internal strife. As early as 1965, in Lusaka, a rival movement was set up by the defectors, the Mozambique Revolutionary Committee (COREMO), initially led by Adelino Gwambe and after by Paulo Guname. In 1968, with the second congress, the Makonde's dissidence – spurred by Lázaro Kawandame – exacerbated, resulting in a violent attack directed at the movement's leadership. Finally, around that time, a broad group of students from the Mozambique Institute – founded by the FRELIMO in Tanzania with the purpose of providing an education to their elites – openly rebelled against its discipline and ideological orientation, garnering the support of their fellow students in the United States.

Lacking both the barbaric violence that marked the outbreak of war in Angola, and the real and close threat of armed struggle in Guinea, the first phase of the war in Mozambique, which lasted until the end of the sixties, was marked by containment and was not a significant cause of concern for the Lisbon authorities. Internal strife in the nationalist movement, the FRELIMO's low intensity military activity, away from the urban centres where the majority of the white population was concentrated, geographic and political closeness to the white regimes of southern Africa and the time for mobilising and preparing that the Portuguese forces had enjoyed gave the colonial authorities a certain measure of confidence.

The end of the sixties and the early seventies saw, however, a threefold change in the unfolding of war in Mozambique. A change that affected the FRELIMO, Portugal and the conflict's very nature and evolution.

In February of 1969, in Dar es-Salaam and in unexplained circumstances, Eduardo Mondlane was assassinated, victim of the detonation of an explosive package that had been sent to him. After a transitory period in which the FRELIMO was run by a triumvirate formed by Uria Simango, Marcelino dos Santos and Samora Machel, it was the latter who, in May of 1970, took on the leadership of the movement. Under the direction of Samora Machel, the FRELIMO underwent an ideological radicalisation of its revolutionary orientation and a redefinition of political leadership: the leading authority was now also the movement's military commander, which reinforced a strategy of popular war of Maoist inspiration.

Shortly before, in March of 1970, a new military commander had arrived in Mozambique, General Kaúlza de Arriaga; his arrival marked the implementation of a new military strategy. Until that point, Portugal had carried out an essentially defensive strategy. This latter avoided major confrontations, developed psychological action operations with the populations, and endeavoured to exploit ethnic rivalries between the movement's leadership, from the south, and the populations that inhabited the guerrilla zones, to the north of the territory. With the arrival of Kaúlza de Arriaga, Portugal evolved into an openly offensive strategy. The Operation Gordian Knot and the question of Cabora Bassa (Cahora Bassa, as it was officially renamed after the independence) were two key moments of that new strategy.

From 1970, with the new military leaders, Kaúlza de Arriaga and Samora Machel, the war enters a new phase. Until then, the guerrilla was stabilised to the north, on the Makonde Plateau and in Nyassa. The FRELIMO had attempted to penetrate Tete from Zambia, but did not succeed. And yet, for the new military commander, inspired by Vietnam-based US doctrines, it did not suffice to stabilise the guerrilla and restrain the movement. It was necessary to defeat the guerrilla and to expunge the FRELIMO. Samora's objectives were, conversely, the same: to defeat the Portuguese army and put an end to colonialism.

A short while after his arrival to Mozambique, Kaúlza begins to devise a colossal operation, meant to be definitive, against FRELIMO: the Operation Gordian Knot.[58] Launched in May of 1970, it combined ground and air resources, employed a new tactic and aimed at the "cleansing" of the zones liberated by the guerrilla to the north of Mozambique. This gigantic military operation was supposed to be complemented by a political-military operation whose purpose was to ensure security in the northern Tanzanian border through the installation of political-military structures and villages: the Operation Frontier.

The Operation Gordian Knot enjoyed short-term success. It managed to dismantle FRELIMO's support bases on the Makonde Plateau, to disband the guerrilla fighters and to substantially reduce military activity in the north of the territory. All this was accomplished in a short time and with little casualties on the side of the Portuguese military forces.

The medium and long-term outcome of the operation was different, however. Not only did it not politically destroy FRELIMO, it also did not prevent the return and reassembly of the guerrilla fighters sheltered at the Tanzanian sanctuary. First, to northern Mozambique and then, progressively, to the Tete zone, where the FRELIMO elected a new strategic goal for the guerrilla: the Cahora Bassa Dam.

Ever since 1966, the Portuguese government had been working on a great hydroelectric project that would have a huge economic and political impact: the construction of the Cahora Bassa on the Zambeze River.[59] The project had, in fact, an immediate and overt economic objective, but also, in the medium-run and more subtly, a political purpose. It was meant not only to produce electric energy intended for South Africa – with a capacity to meet 70% of Johannesburg's needs – but also to irrigate around 400,000 hectares of new lands intended for agriculture. The investment was high and the capital involved was 20% South African and 80% appurtenant to an international consortium – ZAMCO – with French, English, German and Italian participation. Hydroelectric production was meant to be an important source of revenue, from the seventies, but, above all, the huge infrastructure should eventually boost pre-existing ones – ports and railways – allowing for greater penetration and Portuguese integration in southern African economy. This, in itself, was a political desideratum. In fact, the political strengthening of relations with Southern African white powers and the involvement of European capitals in the enterprise had an implicit political meaning: the commitment to the Portuguese project. The creation of settlements and the creation of farming companies made possible by the new irrigated lands was a further objective. It was envisaged that the establishment of settlers would amount to one million. It was expected to reinforce white presence and integrate it in the region and change the pattern of colonisation, containing the spreading of nationalism. It had the opposite effect, however.

The FRELIMO perceived this project as a direct affront and made Cahora Bassa a strategic goal for the guerrilla. Lacking the necessary military capacity to carry out a direct attack on the building site, it focussed on sabotage actions that hindered the construction of the dam: planting landmines on roads, attacks on the railway and the use of harassment of the populations. From 1971, the guerrilla – which the FRELIMO sought to expand across the territory – gathered momentum. It attempted to penetrate into Zambezia and gradually moved to the south. In 1972, it crossed the Zambeze and infiltrated into southern Tete, into Manica and Sofala. In 1973, moreover, it threatened the strategic axis that was the road between Beira and Rhodesia, and came dangerously close to the white settlements, which then began to feel genuinely at risk. Kaúlza responded with the special troops and the Africanisation of war. Around this time, in any case, already more than 50% of the Portuguese troops were African.

The final years of the war were marked by clashing events, all with a political impact on the picture of war in Mozambique. On the one hand, the strained relations with the Church: the expulsion of white priests, the imprisonment and expulsion of protestant missionaries, the conflict with the Bishops of Beira and Nampula and, finally, Father Hastings's recounting of the Wiriamu massacre in the *London Times*. His report had enormous repercussion in the media, especially during an official visit to London by Marcello Caetano. On the other hand, the growing discontent of the white settlers who, when faced with the real threat, inexplicably reacted against the Portuguese Armed Forces and protested loudly in front of the officers' mess in Beira. This was in January of 1974. Finally, something which, in the framework of Portuguese colonies, was a feature of Mozambique alone: an attempt at Rhodesian-style white independence solutions orchestrated by Jorge Jardim.[60] Naturally, it met with no success.

During the ten years of the war, Portugal kept in Mozambique an average force of 31,000 soldiers, having registered a total of 1,569 men killed in action, which means an annual average of 157 killed in action, the highest among the three operational theatres.[61]

In 1971, the FRELIMO claimed, at the United Nations, the control of a third of the Mozambique territory and the liberation of one million Mozambicans.[62] This is most likely an unrealistic estimate, as the political declarations of conflicting parties are bound to be. Around the same time, an estimate by the missionary priests – above suspicion in this matter – pointed to around 100,000 people under the control of the movement.[63]

In short, if, in Angola, the liberation struggle is a disjointed business conducted by three contending movements, and Portugal has military control over the situation, and in Guinea, on the contrary, it is conducted by a single movement endowed with a strong leadership, and the situation is critical, at the military level, for the Portuguese forces, Mozambique's is, in every respect, a half-way experience between those of Angola and Guinea.[64]

This means that in Angola, Guinea and Mozambique three distinct and separate wars were fought. In the three operational theatres, conditions in the ground, employed tactics and military outcomes, both for the Portuguese government and for the liberation movements, were greatly contrasting. However, the political effect is that of a single African war: "overseas war" for some, "liberation war" for others.

Granted, there were common denominators among the liberation movements: a degree of ideological cohesion, economic and political programmes and, above all, the pervading goals of national independence and strategies for the post-colonial states. Also for Portugal, the war effort as a whole was a burden on Portuguese society. And yet, it was really three national conflicts, three totally different decolonisation wars. Why, then, this idea of separateness, when reality pointed to the opposite direction? Because, in the midst of the Cold War, that unification idea was not devoid of political usefulness, both to Salazar and to the nationalist movements: to the latter, as a means to show that they were all engaged in the same fight against colonialism and imperialism; to the former, on the contrary, to prove the Western world that he was championing a truceless crusade against international communism.

War assessment and the political fate of the regime

Notwithstanding the differences among the three conflicts, it is possible to outline an overall assessment, from the point of view of Portugal's interests, of the three cases, and to find the general lines of a common strategy for the three operational theatres.

In the course of the thirteen drawn-out years of the war, Portugal employed an average contingent of 105,000 active duty soldiers in the three theatres. The war exacted a toll of 8,831 dead, of which 4,280 killed in action, which means an annual average of 384 soldiers killed in action. Compared to other contemporary guerrilla warfare scenarios, these numbers are relatively modest. The overall proportion of the daily death rate to one thousand soldiers stands, for Portugal's three decolonisation conflicts, at 0.0075, whereas Algeria registered 0.0107 and Vietnam 0.0365.[65]

Despite the specific features of each of the theatres of war, Portugal's military strategy was, on the whole, in the first half of the sixties, a defensive strategy, based on the square division and directed to three chief objectives: first, to control the main towns and villages; second, to keep communications open and fully functional; third, to launch special operations tasked to carry out attacks on the liberation movements' troop concentrations. The first two missions were carried out by regular troops organised in a square division, which, in their occupation zone, undertake not only military actions, but also "psychological actions" to meet the needs of the population. The third mission was assigned to a team of counter-guerrilla special forces: commandos, *fuzileiros* and paratroopers, and other special forces. From the mid-sixties and, especially, throughout the seventies, the Portuguese strategy becomes more politically, diplomatically and militarily offensive. Politically, it banks on a reinforcement of white colonisation and the region's economic development and integration. Diplomatically, on gathering support and strengthening relations with the white regimes of Southern Africa. Militarily, particularly in Guinea and Mozambique, on increasing the number of intervention special operations and on the Africanisation of war.

Regardless of the individual fate of the three conflicts, this much was certain: the war, that kind of war, could not be settled in military terms.

The furtherance of the conflicts placed the war in the centre of relations between the military and the regime. Between 1961 and 1968, the military institution appeared to unreservedly support the war and, alongside it, the obdurate defence of the colonies and the regime itself. However, from 1968, with Salazar's retirement and Marcelo Caetano's rise to power, the situation begins to change. The perpetuation of the war with neither a military nor a political solution in sight elicits the voicing of the first dissonant positions within the Armed Forces. From 1972, a number of generals, such as Spínola and Costa Gomes, advise against the maintenance of the war effort, which they deemed unfeasible. Among the military, a realisation spreads that there is no military solution to the war. The war is merely buying the government time to devise a political solution. Discontent grows among the captains. The war issue was irrevocably politicised and would ultimately be the driving force behind the Armed Forces Movement which, on the 25th of April of 1974, would overthrow the regime and clear the path, in Portugal, for a transition to democracy.

CHAPTER
8

From War Campaigns to Peacekeeping Operations

Portuguese democracy between Africa and Europe

The revolution of the 25th of April of 1974, as well as the process of transition to democracy which was then initiated, brought about not only a domestic change at the level of political institutions, but also a profound redefinition of Portugal's international positioning and of Portuguese foreign policy. In accordance, naturally, with the spirit of the programme of the Armed Forces Movement (MFA), condensed in the celebrated formula: "democratisation, decolonisation, development". Now, although the MFA's programme ensured the fulfilment of all international commitments taken on by Portugal, it became clear that these two simple principles – democratisation and decolonisation – would require a thorough reinterpretation of those commitments and an in-depth revision of the external orientation of the Portuguese State.

As early as 1974, negotiations were opened with a view to handle the decolonisation of the African colonial territories. The decolonisation was, in fact, the first great challenge of the new regime's foreign policy.[1] There was no consensus on the subject and the truth is, behind the curtain of power, there was a clash of dissonant ideological and political conceptions as to how Portuguese decolonisation should be accomplished. A first trend, heir to Spínola's proposal in his book *Portugal and the Future*, insisted on the federative theory; a second trend, inspired by Melo Antunes, sought the constitution of a neutral, non-aligned and third-world axis; finally, Vasco Gonçalves endorsed a pro-Soviet trend. Politically, these ideological nuances were divided into two essential positions: the first contended that self-determination did not automatically entail independence and campaigned for Portuguese sovereignty until the holding of a referendum which ought to take place under international supervision and decide the fate of the colonial territories; the second, on the other hand, pleaded the identity between self-determination and independence he championed for the immediate transfer of power to the liberation movements, as legitimate representatives of the colonial peoples. After an intricate process, with important consequences for domestic policy itself, the second position won. While, on the ground, a ceasefire was called, the chancelleries initiated the first diplomatic negotiations. Guinea-Bissau, which had already and unilaterally declared its Independence in 1973, was the first country to be internationally recognised by the former colonial power. This was in August of 1974. Between that August and November of 1975, the same process of transfer of power to the liberation movements took place in every former Portuguese colony.

In parallel to the decolonisation dynamics, Portugal was also busy establishing diplomatic relations with the Soviet Union and the Eastern and Third World countries, with the exception of Albania and China, with whom the process, due to the question of Macau, ran into a number of difficulties which would only be settled in 1979.[2]

However, decolonisation and diplomatic openness to the world, per se, were insufficient to define the new external orientations of Portuguese democracy. Instead, under the overt and noisy political clashes concomitant with the process of internal democratisation, another battle was being fought, invisible and silent, over the country's international situation and the strategic options of Portuguese foreign policy.

Between April of 1974 and January of 1986, Portuguese foreign policy oscillated between two basic orientations which also mark two distinct phases: the transition to democracy, corresponding to the pre-constitutional period and ruled by the revolutionary process; and the democratic consolidation, corresponding to the constitutional period and marked by the institutionalisation and stabilisation of the democratic regime.

The pre-constitutional period was punctuated by struggle over the country's external options, by the exercise of parallel diplomacy and, consequently, by an overall uncertainty regarding foreign policy. Despite the struggle, the indecisiveness and the pervading uncertainty that permeated the provisional governments, especially those in which the military had a more preponderant role, the general orientation of Portuguese foreign policy tended towards a third-worldist option and the pursuance of privileged relations with the new countries arising from the Portuguese decolonisation. In a sense, this was the last avatar – albeit now a socialising one – of Salazar's cherished thesis of Portugal's "African vocation". Regardless of that vague and diffuse orientation towards neutralism and the temptation of non-alignment during the process of transition to democracy, neither did the 25th of April, nor the 28th of September, alter the orientation of the Atlantic axis and the military relations with NATO. Only the 11th of March of 1975 ultimately disrupted those relations at the political and military level. And the reason for this was simple enough. In the midst of the Cold War, a NATO member with communist ministers in the government was not in the least reassuring for the Alliance's partners. And Portugal found itself temporarily excluded from the Eurogroup and all classified military information. On the other hand, in Portugal, notwithstanding the criticism directed at NATO and the North American "imperialism" by the forces further to the left, not even the Portuguese Communist Party (PCP) demanded Portugal's exit from the Atlantic Alliance.[3]

The constitutional period which began precisely with the first constitutional government was marked by the clarification of Portuguese foreign policy and a univocal and rigorous definition of Portugal's external positioning. A Portugal which thenceforth committed to its status as a Western country, at once European and Atlantic. If, to these orientations, we add the reinstatement and normalisation of post-colonial relations with the new countries emanating from the Portuguese decolonisation, we have before us the three main axis of the foreign policy of Portuguese democracy.

Between 1976 and 1986, the Atlantic vector and Atlanticism prevailed still, having played an important role, at the level not only of the country's international positioning, but also of internal stabilisation and democratic consolidation.[4]

Thus, the Atlantic vector meant, for Portugal, reconciliation with tradition and the endurance of the historical features of its foreign policy. Bilaterally, that Atlanticism substantiated in the strengthening of diplomatic relations with the United States and the renewal of the Lajes Agreements in 1979 and 1983.[5]

In 1977, a joint declaration by Portugal and the US states the willingness of both governments to conclude negotiations over the facilities in the Azores base, which eventually happened in June of 1979, and whose outcome was the extension of the lease on the base until 1983. The 1979 agreement also pledged the provision of economic compensation, which materialised in an annual financial aid of 20 million dollars granted to the Autonomous Region of the Azores. With the end of the military embargo on Portugal, the economic aid was eventually accompanied by a military grant: around 60 million dollars for defence.

The aggravation of the situation in the Middle East, the overthrow of the shah of Iran and the Soviet invasion of Afghanistan, in the early eighties, pressed the need to reinforce Western presence. It was within this framework of strategic reappraisal of the Azores base that the 1983 agreement was signed.

The use of facilities in the Lajes base was extended until 1991 and, as compensation, Portugal was once again the recipient of economic and military aid. At the military level, Portugal was awarded a 517.5 million dollar grant and a 412.5 million dollar loan for eight years. As economic compensation, it was given a 75 million dollar loan and a 320 million dollar donation. In this scope and imbued with the same spirit, the Luso-American Development Foundation (FLAD) was created.[6]

At the multilateral level, the Atlantic vector translated into the maintenance and reinforcement of Portugal's position in the Atlantic Alliance, as well as the redefinition and renewal of Portuguese commitment to the military engagements with NATO, which the war effort in Africa had forced to put on hold since the sixties. With respect to the Army, this reassertion of the NATO commitments was substantiated by the creation of the Independent Mixed Brigade (BAI), since then converted into the Air Transport Brigade (BAT), which replaced and, to some extent, reactivated, the former Independent Division of the Army, which essentially furthered the goals of its predecessor in the military missions of NATO's southern flank. With regard to the Navy and the Air Force, patrol missions carried out in the scope of IBERLAND were reinforced. IBERLAND's command was upgraded to command-in-chief – CINCIBERLAND – and the position was entrusted to a Portuguese officer.[7]

The "European option" is, however, the great novelty of post-25th of April foreign policy and, from 1986, a topmost priority in democratic Portugal's external orientation.[8]

Having overcome anti-European resistance which stemmed, first from the Africanist option of the authoritarian regime, and afterwards from the third-worldist temptation of the revolutionary period, Portugal committed entirely, from 1976, to the "European option". The latter was no longer spurred by merely economic and pragmatic motivations as had been the case by the end of the authoritarian regime. Now, it represented a political project and the future of the country. Portugal's veering towards the European integration process began precisely in that year of 1976, with its accession to the Council of Europe and the signing of the additional protocols to the 1972 Trade Agreement between Portugal and the EEC, which indeed made up, to a large extent, the pre-accession process.

After a successful round of negotiations involving the European capitals, between September of 1976 and February of 1977, the

I Constitutional Government made a formal request for Portugal's EEC membership. This was in March of 1977. With the formal request for membership, all the dithering surrounding the formula for Portugal's accession – either its pre-accession status or the so-called "privileged association" – was once and for all overcome and the "European option" effectively fulfilled. It was a strategic option marking decisively the country's fate.

Two primary objectives led the government to adopt this strategy: in the first place, the democratic consolidation that Portugal's accession to the Community ensured; secondly, the economic modernisation and development that the EEC structural funds would assist in accomplishing.

The formal request for membership was followed by a long and intricate negotiation process that extended for almost a decade. The process culminated in June of 1985 with the signing of the Treaty of Accession to the EEC by Portugal. On the 1st of January of 1986, Portugal became a full member of the European Community.[9] Thenceforth the "European option" became the main strategic axis of Portugal's foreign policy and the topmost priority of the national interest. The European and Atlantic vectors were joined by a third, post-colonial vector: the development of relations and ties of friendship and cooperation with the new Portuguese-speaking countries. After the first moment that immediately followed the decolonisation processes, this third vector will represent a major concern, not only for the successive governments, but also for the Presidents of the Republic, who did not spare diplomatic efforts to promote to the normalisation and improvement of relations with the new Portuguese-Speaking African countries (PALOP). This process, which had been initiated on a bilateral basis, finally peaks in 1996 with the multilateral institutionalisation of the Community of Portuguese-Speaking Countries (CPLP).[10]

Therefore, without effecting any change in its international standing, Portugal altered, however, its strategic priorities. Historically, it had regarded itself as an Atlantic and colonial country that, whenever the weight of the maritime vector proved excessive, sought a counterbalance in Europe. From that moment, it has stood as a European country that preserves and endeavours to capitalise on its Atlantic position and its post-colonial relations.[11]

Europeanisation, Atlanticism and post-colonial relations constituted thenceforth the essential orientations and the strategic priorities of Portugal's external positioning and international presence. It was against this background and in accordance with these priorities that external defence commitments and the new international missions of the Portuguese Armed Forces were developed: military commitments with NATO, in the Atlantic framework; involvement with the Western European Union, at a first moment, and, afterwards, with the Common Security and Defence Policy, in the European context; and technical-military cooperation with the PALOP, in the post-colonial framework. In the scope of these external commitments and international missions, and since the post-Cold War period, the presence of the Armed Forces in humanitarian crisis management missions and peacekeeping operations will be called to stand out.

The post-Cold War period: Atlanticism, Europeanisation and post-colonial relations

The end of the Cold War, the fall of the Berlin Wall and of the Soviet Union, the dissolution of the Eastern Bloc and the Warsaw Pact, German reunification and post-Soviet drift in Central and Eastern Europe, as well as the persistence of the United States as the sole global power, led to the emergence, since 1989, of a new and unipolar international order and of a new framework of world security.

The new international strategic environment, on the one hand, and the country's new external standing, on the other, shaped the evolution of a new framework for Portugal's international position. The traditional Atlanticism is joined by Europeanisation, and the bilateral and multilateral dimensions of the diplomatic relations with the US and NATO, respectively, are joined by Portugal's new status as a member of the European Union.

From that moment, a controversy between two different trends flared up in the midst of Portuguese political debate and strategic thought. The first, which might be called "Atlanticist", stems from a Eurosceptic outlook and contends that Portugal, occupying a peripheral position in Europe, should not commit itself to Europe. For this trend, the area of prime strategic interest for Portugal was the Atlantic and the return to Portuguese-speaking Africa. That should be therefore the priority of the Portuguese Armed Forces. The second trend, which we may call "Europeanist", on the contrary, argues that, in the complex framework of the post-Cold War order, the border of Portuguese security does not coincide with its national border and that, as such, geographic periphery should be counterbalanced by political centrality. Consequently, not only was Europe, for reasons of regional security, an area of national interest for Portugal, it should constitute top priority for the engagement of Portuguese military forces.

Since the end of the Cold War and throughout the nineties, Portugal's stance in this matter evolved from a traditional, strictly Atlanticist position, to a more Europeanist – indeed Euro-Atlanticist – position.

The European vector undoubtedly represents the great novelty of democracy's external orientation. A member of the European Community since 1986, signatory to the Single European Act, Portugal immediately began to take part in the European Political Cooperation. The Treaty on European Union, signed in Maastricht in 1991, unified under the same juridical and institutional framework the European Economic Union and the European Political Cooperation, outlining a European Union built on a three-pillar structure, whose second pillar instituted the Common Foreign and Security Policy (CFSP). The CFSP was given the same institutional framework as the community pillar, although endowed with different rules and procedures and a decision-making process based essentially on intergovernmentalism. It acquired new instruments besides the common positions, and adopted joint actions. Above all, it broke a European integration taboo that ensued from the failure of the European Defence Community and, for the first time since 1954, it included European defence issues in its scope of application.[12] On the other hand, the Western European Union (WEU), founded by the Treaty of Brussels in 1948, amended and enlarged in 1954 and a "sleeping beauty" all throughout the Cold War, was reactivated in 1984, gaining a fresh impetus and a new international role from 1991. In fact, the Treaty on European Union deemed the WEU an integral part of EU's development and it was the WEU, according to the Treaty, the organisation responsible for the execution of EU actions that involved military implications. To these provisions, laid out in the Treaty, was added an Annex Declaration by the WEU member states concerning its role and relations with the EU and NATO. Whereas, in the Treaty, a European perspective seemed to prevail – with the subordination of the WEU to the EU and, to that end, the creation of a political-military operational apparatus – in the Annex Declaration an Atlantic perspective held sway, with the emphasis on WEU's relationship with NATO. Ambiguity was the price to pay for a hard-to-come-by compromise between the two perspectives on the question of European security: the Atlanticist perspective, which sought a defence solution within the framework of NATO, and the Europeanist perspective, which strove, beyond NATO, for an autonomous defence capacity for Europe. Was WEU the armed wing of the European Union or NATO's European cornerstone? The core issue was the definition of what, at the time, was called the European

Security and Defence Identity (ESDI), an issue that was never indeed solved.[13]

As a member of the EU, Portugal took part from the first moment in the creation of the CFSP. However, because it did not sign the Treaty of Brussels in 1948 and was naturally withdrawn from its amendment in 1954, only after its accession to the European Community was the Portuguese admission considered. The treaty of accession whereby Portugal, together with Spain, joined the WEU, was signed on the 14th of November of 1988 and ratified on the 27th of March of 1990. From then on, Portugal was a full member of the WEU and became fully present in the sphere of defence of the EU.

However, due to WEU's political and military limitations and technical and operational uncertainties, which rendered it unable to face autonomously possible regional threats, a number of more Europeanist member states decided for the creation of alternative European military forces under the shape of a sort of "enhanced cooperation". In 1991, the French–German idea of creating a hard core of joint military units is launched: the Eurocorps. The Eurocorps is born in 1993, an offspring of France, Germany, Spain, Belgium and Luxembourg. Faithful to the Atlantic tradition, Portugal saw in the Eurocorps a strictly continental initiative which, moreover, could jeopardise the transatlantic relationship. Thus, it remained on the sidelines and did not join in.

In terms of strategic orientation, Portugal preserved still an Atlanticist position and was radically opposed to any continental rapprochement, especially regarding security and defence. However, as its experience of European integration deepened, and particularly after its first presidency of the European Union, in 1992,[14] Portugal travelled a path of Europeanisation in its foreign policy,[15] which brought it closer to a Euro-Atlantic position. This European evolution of Portuguese strategic orientation became especially prominent from 1995 and was clearly reinforced during the Socialist Party governments, between 1996 and 2000.

In October of 1994, a Portuguese ambassador, José Cutileiro, was appointed Secretary-General of the WEU. It was the first time that Portugal was assigned such a high position in a European defence organisation, which also bore testimony to the pro-European evolution of the Portuguese position.

At the Lisbon Summit, in 1995, the WEU decided to create, in addition to the existing political-military structures, a permanent military staff in Brussels, which, despite not being equipped with permanent armed forces, now enjoyed a number of operational units that included two forces: a light rapid reaction force – European Rapid

Operational Force (Eurofor), with headquarters in Florence; and a naval force – European Maritime Force (Euromarfor), in the Mediterranean. With the participation of France, Italy, Spain and Portugal, these Mediterranean forces had a rotating command and, in case of need, may be composed or reinforced by military units deployed from national forces. The option to join these Mediterranean forces with the southern countries, rather than the Atlantic option to join the United Kingdom and the Netherlands, who were forming other forces, confirms and reinforces the sign of a growing Europeanisation of Portugal's external orientation in terms of defence.

In 1995, one other clear indication: Portugal became "permanent observer" of the Eurocorps. As to the ESDI, Portugal took on, from that point, a clearly favourable position. It supported the concept of Combined Joined Task Forces (CJTF) and the use of NATO military capabilities by the WEU, namely in the so-called Petersberg tasks. Moreover, in defiance of what had always been its traditional strategic position, Portugal accepted and supported the long-term integration of the WEU in the European Union.[16]

This position, defended by Portugal at the intergovernmental conference for the revision of the EU treaties indelibly marks a historical reorientation in the country's strategic options. It is, for Portugal, the end of the anti-European Atlanticism and the affirmation of its transatlantic link as a European country.[17]

The Treaty of Amsterdam did not introduce substantial changes to the CFSP, namely in terms of defence. It simplified the decision-making processes, but preserved its essentially intergovernmental character. It added the common strategies to the CFSP instruments. It brought in an institutional innovation, with the establishment of a High Representative of the Union for Foreign Affairs and Foreign Policy and a Planning and Early Warning Unit, but the same did not happen in the defence sector: the EU and the WEU remain institutionally distinct, even if the treaty incorporated the Petersberg tasks – WEU tasks – in the framework of the EU. It was an operational, but not an institutional convergence.

At the level of security and defence, however, the big change, as far Europe was concerned, had to wait for the British–French Saint Malo Summit in 1998, when, for the first time, the United Kingdom agreed to the creation of an autonomous defence capability within the EU. The first step had been taken towards the establishment of a European Security and Defence Policy (ESDP), which would be set out in the Treaty of Nice, in December of 2000.

After Nice, the WEU was incorporated into the EU, which absorbed its institutions and capabilities and set in motion the institutionalisation of the ESDP. The incorporation took place at two levels: institutions and military capabilities. At the level of the institutions, three new permanent bodies were created within the EU, as counterparts to NATO's or the extinct WEU's political-military structures: the Political and Security Committee (PSC), the Military Committee and the Military Staff. The European Union now had – an absolute novelty until then – a military dimension. At the level of capabilities, the debate had been going on since Saint Malo and took on a political shape with the decision, by the Helsinki European Council, to create a Rapid Reaction Force (RRF) able to deploy 60,000 men within 60 days and sustain such a force on the ground for 12 months, and which should be ready for operations by the end of 2003. Their goal was to carry out Petersberg tasks.[18]

Ever since the beginning of the process, following Saint Malo, Portugal participated in all the European summits – Cologne, Helsinki, Santa Maria da Feira and Nice, and supported the creation of the ESDP. Furthermore, Portugal announced, at the Capabilities Commitment Conference in Brussels, in September of 2000, and in the framework of the Helsinki Headline Goal, that it would make available to the RRF a contingent of 1,000 soldiers.

The launching of the ESDP was the European response to its manifest capability shortfalls in the Kosovo crisis. Its more significant development, however, was precipitated by the quick and profound change in the international strategic environment after September 11th.

Since its official establishment by the Treaty of Nice, in 2000, until the Treaty of Lisbon, in 2010, the European Security and Defence Policy (ESDP) stood as one of the most dynamic areas of European integration at its various levels.

At the institutional level, the ESDP set in motion the institutions provided for in the Treaty, endowing the European Union with the capacity to respond to decisions with military implications.

At the level of capabilities, after the Helsinki Headline Goal, it approved a new 2010 Headline Goal with a view to creating a European rapid response capability, including the training of Battlegroups, in operation since 2007.

To the advancements already in progress at these two levels, a third one was added at the conceptual level: the European Union approved, for the first time in 2003, an unprecedented document defining common threats and risks and also a European strategic concept and perspective on the Union's external action. Five years later, in 2008, the European Security Strategy was reviewed and updated in the light of the evolution of the international scene.

At the operational level, the European Union began to carry out civilian and military operations on the ground. Initially, in coordination with NATO, under the Berlin Plus agreement. Later on, autonomously, in several theatres of operations. Since its beginning, in 2000, and until the Treaty of Lisbon came into force, in December of 2009, the ESDP had promoted 22 missions and deployed around 7,000 soldiers.[19]

The Treaty of Lisbon, in 2007, consolidated and developed the work undertaken throughout a decade of European Security and Defence Policy. In the first place, with a change in nomenclature that had great political significance: it became the Common Security and Defence Policy (CSDP). In the second place, with the introduction of two important solidarity clauses: the first, a mutual defence clause, after the model of the article V of the North Atlantic Treaty; the second, in the event of natural disasters or terrorist attacks. Thirdly, with a clearer and more exhaustive definition of the targets for the Union's civilian and military capabilities. And, finally, the introduction of the mechanism of Enhanced Cooperation in the scope of defence and the creation of the Permanent Structured Cooperation, the most challenging instrument, yet the one most likely to contribute to a reinforcement of a European defence policy.[20]

Portugal was always a part of the whole process and invariably favourable to all the developments that led from the ESDP to the CSDP. It participated in all the Union's military missions and was present, from the very beginning, in every step of the development of its military capabilities.

Now, if the European vector represented the great novelty of Democracy's foreign and defence policy, the Atlantic vector, on the contrary, constituted its element of permanence and reunion with the historical tradition, both at the bilateral and the multilateral levels.

At the bilateral level, Portugal's relations with the United States were marked both by a readjustment in US security and defence policies for the post-Cold War era and by Portugal's new European dimension, and finally led, in 1995, to the signing of a new Agreement on Cooperation and Defence, which marked the beginning of a new model of relationship between Washington and Lisbon.

The end of the Cold War brought about a huge change in US security policy, which in the early nineties entered a period of lower investment in the defence sector, at least in its traditional premises. Not only were military bases deactivated all across the North American territory, as US presence in the European theatre was diminished. And, following George Bush's so-called Regional Defence Strategy and Bill Clinton's Bottom-Up Review, the concept of "advanced defence" based on which, since the second post-war period, a worldwide network of North American military bases had been built. US military presence abroad had turned to other priorities: smaller yet more flexible, mobile and effective forces. From that point onwards, the US deactivated, totally or partly, many of their military bases abroad. At the same time, the traditional philosophy that governed the foreign assistance programmes began to change, reducing military aid in favour of economic and humanitarian assistance programmes.

This new atmosphere had an immediate impact on US relations with Portugal, namely concerning the Azores base. This, together with economic development, social stability and the consolidation of democracy, which are confirmed with Portugal's accession to the EC, weighed decisively in the negotiations that were opened for the new agreement in 1991.

The US attempt to strategically play down the Lajes Base and the Portuguese insistence on the model of financial compensation and the extension of the military aid hampered bilateral negotiations and led to a deadlock in 1992. From 1993, positions became more flexible and, in the face of European integration and the country's level of development, Portuguese foreign policy not only was forced to relinquish the traditional "facilities vs. compensation", but, above all, adopted a new way of looking at transatlantic relations. In Portugal's view, the relationship with the single global power transcended the specific question of the Lajes Base, which paved the way for an agreement of a general nature. On the 1st of June of 1995, the ministers of Foreign Affairs of both countries signed, in Lisbon, the Agreement on Cooperation and Defence, but also a technical and labour agreement aimed at the Lajes Base employees.

The new Agreement on Cooperation and Defence did not define specific programmes, but identified the various areas of cooperation: military and defence, scientific and technological and in the domain of economic and commercial relations.[21]

At the multilateral level, Portugal's presence in NATO was still marked by the maintenance and reinforcement of its political-military commitments, but also by the changes and adjustments of the Alliance in the post-Cold War period. And, in parallel, by Portugal's accession to the Community, by its accession to the WEU in 1990 and by its later engagement with the ESDI.

Created for the Cold War and designed against the Soviet threat, NATO not only survived the death of its enemy – the reason for its existence – but also persisted in a world for which it had not been created. In the early nineties, the US Secretary of State James Baker insightfully pinpointed NATO's quandary: "either extinction or transformation, maintenance of the status quo was impossible". NATO chose transformation and initiated a complex reform process in order to adapt to new functions and new missions. Reforms both at the external and internal levels, with a political and military dimension.[22]

At the external level, NATO's agenda included two major problems: the question of the enlargement to the east and the preservation of the transatlantic link.

The Alliance's first political challenge was the eastern enlargement, a symbolic token of the victory in the Cold War and, at the same time, a way to fill the strategic void left by the dissolution of the Warsaw Pact and a security factor for post-communist Europe.

Confrontation gave way to cooperation. The old enemies became partners and allies. In that process, NATO developed a three-phase initiative: the establishment of a cooperation structure, launched at the Rome Summit of 1991, the North Atlantic Cooperation Council; the offer of a strategic partnership proposed at the Brussels Summit of 1994, the Partnership for Peace; and, finally, the enlargement, decided in Madrid in 1997, and granted to three countries – Poland, Hungary and the Czech Republic. Russia was neither forgotten nor excluded from the new architecture of European security. It was even the object of a special rapprochement substantiated in the Paris Founding Act of 1997, which associated Russia to the Alliance through the Euro-Atlantic Partnership Council.

Throughout the external reform and the enlargement process, Portugal consistently showed a spirit of convergence and openness to the evolution of the Atlantic Alliance. It participated in the establishment of every new institution, it developed cooperation relations in the scope of defence with the Central and Eastern European countries, in the framework of the Partnership for Peace, and actively supported NATO's enlargement. At the Madrid Summit, three new members were invited to join NATO, and Portugal supported, along with other European Countries, the 3+2 formula, which meant a wider enlargement not only to Poland, Czech Republic and Hungary, but also to Slovenia and Romania.[23]

Since then, the process of enlargement has not ceased and there have been no break-ups. At the Washington Summit, in 1999, the Alliance decided that it should not hamper the entry of new members and defined the principles of the method to adopt: it would welcome candidates who fulfilled the political and military requirements, that is, who were willing to further the founding political principles of the Alliance – democracy, individual liberty and the rule of law – and whose military capabilities were in a position to contribute to collective security and international stability. In 2009, the enlargement extended to Albania and Croatia. The Alliance now included 28 members. In the spirit of these principles, Portugal was favourable and consistently supported the enlargement process.[24]

The other item in NATO's external agenda concerned the relationship between the two pillars of the Alliance – the American and the European – that is, with the transatlantic relation.

Throughout the length of the Cold War, the transatlantic link had ensured European security and even international peace and stability. In the post-war period, their maintenance continued to be perceived as an essential condition for European security. The evolution towards a unipolar international system with the US as the sole global power, and the EU's institutionalisation of the CFSP, reactivation of the WEU and setting in motion of the ESDI, however, pressed for a reassessment of the transatlantic relation.[25]

The Atlantic Alliance immediately reinterpreted the ESDI, not as "an armed wing of the European Union", but rather as "NATO's European pillar". And not only, in the following summits and declarations, from 1992 onwards, did it support this notion of European Security and Defence Identity, the Berlin Summit of 1996 also declared the creation of the ESDI within NATO itself. Portugal supported at all times the development of the ESDI and evolved, in this matter, from a strictly Atlanticist to a Euro-Atlantic position.

Although the establishment of the European Security and Defence Identity by the Treaty of Nice and, at long last, the Common Security and Defence Policy by the Treaty of Lisbon constituted first steps in the direction of an autonomous EU defence capacity, afterwards the core problem was not substantially altered. US intervention in Iraq, in 2003, led to the gravest transatlantic crisis after France's withdrawal from NATO's integrated military structure in 1966, and had profound consequences. The crisis was eventually overcome and the transatlantic link re-established. The direction of the Obama administration's policies, in the US, and France's return to NATO's military structure, in Europe, closed that cycle once and for all. But relations between NATO and the European Union, NATO–CSDP in particular, are still far from resolved. Both at the political and the institutional and operational levels. Proof of that is the ambiguousness of the terms of the NATO–EU strategic partnership in the new strategic concept of the Alliance, approved in Lisbon in 2010.

In the course of this period, the Portuguese position never changed. Following a progressively more Europeanist trend, Portugal supported and integrated from the outset the ESDP and the CSDP, and joined in every international mission carried out in that scope. It advocated, however, at all times, a notion of strategic complementarity between NATO and the EU. In terms of internal reform, NATO's agenda included two fundamental issues: the restructuring of military commands and the definition of a new doctrine and new missions.

The restructuring of commands aimed at responding and adjusting NATO to the new strategic scenario. Although it preserved in its structure the two strategic commands, the Allied Command Atlantic and the Allied Command Europe, it undertook reforms in lower-level commands: three regional commands in the SACLAND, two regional commands in the SACEUR and, in within both, other sub-regional commands.

For Portugal, the main goal in this reorganisation of commands was the preservation of *CINCIBERLAND* in Portuguese territory and its permanence at the level of regional command. The question was raised by Spain's entry into NATO's integrated military structure and the decision concerning the establishment of a command in Spanish territory. This was not a bilateral, but a multilateral question in the NATO framework – the definition of responsibilities between the two strategic commands, *SACLAND* and *SACEUR*. In the end, however, it received a diplomatic solution negotiated between Portugal and Spain, a solution presented to the Alliance and later ratified by its military and political bodies.

The new philosophy of the allied commands merely defined the limits of its strategic commands, and not those of the lower-level commands. These limits and their military responsibilities were, in case of need, defined by the strategic commands, with the establishment, in the areas of mutual interest, a *supporting command* and a *supported command*.

In the spirit of this philosophy, Portugal succeeded in reaching a deal that satisfied its objectives entirely. In the first place, it preserved in its territory not only the Iberian-Atlantic command, but also the level of the same command. In the second place, and for the first time in the history of its presence in the Alliance, the Portuguese territory, mainland and Atlantic, was wholly integrated under the same command: the SACLAND. (Ever since the first definition of NATO's military commands, the defence of the Atlantic component of the Portuguese territory had been entrusted to the Atlantic Command, but the defence of the mainland territory was a national responsibility.)

In the third place, it managed to prevent the fulfilment of the Spanish claim to a strategic corridor between Gibraltar and the Canaries under the command of the *SACEUR* and, in particular, of Madrid's sub-regional command. Except for a "bubble" over the Canary Islands, resulting from the compromise that had been reached, and which fell under the command of Madrid, the responsibility for all the Atlantic space to the west of the Guadiana River meridian and to the north of the Tropic of Cancer was assigned to *SACLAND*.

Spain, in its turn, upon entering NATO's integrated military structure, managed right away to obtain a sub-regional command in its territory. It managed, moreover, to keep the whole of its territory, including the Canaries, under the same command – the SACEUR – and to shift westward – from Gibraltar to the Guadiana River – the demarcation line between the Alliance's two strategic commands.[26]

After September 11th, changes in the international strategic environment and new threats and risks continued to compel the Alliance to pursue the development of its transformation process, namely the adjustment of its military structures to the new missions.

After the Washington Summit, in 1999, not only did the NATO command remain in Portugal, it was furthermore transformed and enhanced in one of the Alliance's three regional commands: *CINCIBERLANT* was upgraded to *CINCSOUTHLANT*. In 2003, it is transferred from *SACLANT* to *SACEUR*. And, in 2004, it becomes Joint Force Command Lisbon, an operational and deployable command, able to command Alliance missions in any of NATO's areas of responsibility and certified as a NATO Response Force (NRF). In 2006, it took action and demonstrated its effective capacity at the command of relief operations in at least two occasions: the earthquake in Pakistan and Hurricane Katrina.

In June of 2011, in the scope of the transformation process, and in the wake of the reform of military structures which led to a reduction in staff from 13,000 to 9,000 posts and in the number of Agencies from 14 to 4, and also the closing of several bases, *Joint Command Lisbon* was deactivated. NATO's flag, however, remained in Portuguese territory, whereto *STRIKFORNATO*'s headquarters were relocated from Naples, and which includes US Sixth Fleet.

Finally, the last question concerns the new doctrine, that is, the strategic concept of the Alliance. Ever since the end of the Cold War, NATO drafted three strategic concepts, and each new one sought to respond to the evolution of the international scene, namely the change in the strategic environment, that is, the type of threats and risks faced by international security.

In 1991, in Rome, it was a matter of tackling the end of the Cold War, Germany's reunification and the transition to democracy of East-Central European post-communist drifts. Not long after, the dissolution of the Soviet Union came to change once again the international environment. Rome's "strategic concept" pointed towards a new – not strictly military, but progressively more global and integrated – notion of security. From that moment, besides the traditional "collective defence" missions defined by article 5 of the Washington Treaty, a new type of mission began to find its own space which the Alliance came to approve in the Atlantic Council in Oslo, in 1992: missions of humanitarian assistance and peacekeeping operations.

In 1999, at the 50th anniversary summit in Washington, it was essentially a two-fold reply: first, to the enlargement of the Atlantic Alliance to the old enemies of the Warsaw Pact, which, in the absence of a peace treaty, recorded in Europe's political geography the victory of the West in the Cold War; second, to the Balkan secession conflicts. Washington's strategic concept added to the already accomplished logic of geographic enlargement one other enlargement logic: an expansion, to begin with, of the Alliance's area of intervention; and, secondly, of the Alliance's types of missions. The disappearance of the traditional threat, the emergence of new threats and the proliferation of new risks and of a new type of international conflict had a twofold effect on NATO. On the one hand, it was becoming more and more unlikely the need for the traditional collective defence mission and increasingly more likely the pertinence of the new type of mission, which included, besides crisis management and peace-support operations, fight against the proliferation of weapons of mass destruction, terrorism, drug trafficking and organised crime. On the other, these missions appear to be ever less likely within the traditional area defined by the North Atlantic Treaty and gradually more likely outside this area. An issue which instilled in the Alliance the syndrome of the "out of area or out of business" and which led NATO to approve the revision of the strategic concept that provided for these new missions outside the area and, more than that, without the need for a UN mandate, political consensus within NATO being the sole requirement. In the midst of the crisis in Kosovo, and not without international contestation, the revision of the strategic concept was approved, changing NATO, in the early 21st century, from a regional defence alliance into a tendentially global security organisation. Attending, in Washington, the 50th anniversary summit, in April of 1999, Portugal participated and approved the reform of the Alliance.[27]

In Lisbon, in 2010, it was a matter of tackling the radical changes of the first decade of the 21st century. A change primarily in the international system, marked by the transition from a unipolar world under the Western hegemony of the US to a post-Western and tendentially multipolar world. Secondly, a change in the strategic environment, with the emergence of a new picture of threats and risks. Quite beyond the proliferation of weapons of mass destruction, the failed states or the domestic conflicts of ethnic and religious origin, this new post-

September 11th scenario was dominated by the threat of transnational terrorism. Changes in the international system and in the strategic environment, together with the experience of a decade within an Alliance engaged in new missions outside its area, especially in Afghanistan, account for the new strategic concept adopted in Lisbon.

Portugal hosted and organised the Lisbon Summit in November of 2010. It took part in every phase of the drafting of the concept and supported the agreed strategic options, which it deemed convergent with the national interest.

In the first place, the emphasis on collective defence and the focus on the Euro-Atlantic area. These were of utmost interest to Portugal, not only because, within the framework of unity and indivisibility of the Alliance's territory, the reaffirmation of article 5 reinforced collective defence, but also because by refocusing on the Euro-Atlantic space, Portugal's centrality came out reinforced and the strategic value of the Portuguese territory was re-established.

In the second place, the reassessment of the Alliance's international responsibilities might very well represent, for Portugal, an opportunity to clarify and reaffirm the doctrine of the intervention of the Portuguese armed forces in international missions. A doctrine no longer dictated by exclusively historical or geographic factors but, just as in the Alliance, by international security criteria.

Finally, the Alliance's partnerships were also of strategic interest to Portugal, inasmuch as they constituted an opportunity for the latter to stress its position within the Alliance. First, if the NATO–EU strategic partnership was given leverage, Portugal would see the value of its strategic centrality renewed, and could more easily profit from its Atlantic belonging in the European framework and, at the same time, its European belonging in the Atlantic framework. Secondly, in the establishment, by the Alliance, of future partnerships in areas wherein Portugal relies on historical experience and privileged insight, such as the Magreb, sub-Saharan Africa and the South Atlantic, where it could work as a facilitator, building bridges with regional partners and reinforcing its position within the Alliance.[28]

The third vector of Portugal's international presence in terms of defence and Armed Forces corresponds to the third area of strategic interest for Portugal, which is that of its post-colonial relations. That is, the updating and reinvestment in the African vector. Since the end of the decolonisation process, and after the first difficult years in the bilateral relations between Portugal and its former colonies, these gradually underwent a normalisation process. This normalisation, on the one hand, and the knowledge and experience of the Portuguese Armed Forces in Africa, on the other, opened up the possibility for now using them with a completely different objective: that of technical-military cooperation.

In the framework of the new independent countries, the establishment and training of the military instrument proved to be crucial for the building of the national state. The change from the guerrilla military forces of the liberation movements into conventional armies was not a straightforward process, neither politically nor technically. In the case of Angola, the existence of several liberation movements with their own military forces, which had fought a civil war among themselves, made the process all the more difficult and complex. Now, this was precisely the goal of Portuguese technical-military cooperation: the creation of new Armed Forces. Firstly, non-partisan and subordinated to the democratic political power; subsequently, technically and militarily ready. Cooperation was thus directed, on the one hand, at the legislative and organisational fields of the Armed Forces and, on the other, at the technical field, namely at supporting the restructuring of the different branches: engineering, communications, hydrography, oceanography, maritime signalling, among others.

From 1978, in this context of post-colonial cooperation and with that precise goal in mind, Portugal began to develop relations in the military field with the Portuguese-speaking African Countries (PALOP). During this phase, however, either due to Portugal's domestic situation or to the state of bilateral relations, the military cooperation was limited to *ad hoc* one-off and episodic actions, at the request of this or that Portuguese-speaking African country – a vertical cooperation. The first agreement establishing this type of cooperation was signed as early as in 1978 with Angola: the General Portuguese–Angolan Cooperation Agreement.

It was only from the mid-eighties, however, that democratic normalisation, European integration and the strengthening of post-colonial relations conferred to Portuguese cooperation a transversal coordination and a strategic perspective which, at the military level, translates into the so-called Technical-Military Cooperation Agreements. The first of these was signed in June of 1988, with Cape Verde; the second and the third, in December of the same year, with São Tomé e Príncipe and Mozambique; and the fourth in January of 1989, with Guinea-Bissau.[29] In 1989, the Department of Technical-Military Cooperation is created, subordinated to the Directorate-General of National Defence Policy of the Ministry of Defence and which, from then on, took on the role of studying, planning and monitoring technical-military cooperation with the PALOp. In 1990, the first framework-programmes of technical-military cooperation with Guinea-Bissau, Cape Verde and Mozambique, followed by São Tomé in Príncipe, already in 1991, were approved and set in motion. In Angola, the domestic situations resulting from the civil war held back the whole process and only in 1996 was it possible to sign the General Cooperation Agreement in the Domain of Defence and Military.[30]

By the end of the decade, in 1999, the total number of ongoing projects and sub-projects with the PALOP, in the scope of technical-military cooperation, was 70. On the same date, the number of military involved in the three branches of the Portuguese Armed Forces was 91 and the balance of the PALOP military training throughout the decade, between 1991 and 1999, was very positive: 3,588 trained in Portugal and 8,133 in the African countries.[31] A decade later, in 2010, the total number of projects stood at 49, with the engagement of 256 Portuguese military, and a balance in the PALOP military training of 125 in Portugal and 7,840 in the Portuguese-speaking countries.[32]

The democratic model of the armed forces and the new international missions

In the course of the four decades of democratic regime, and in the scope of these three vectors of strategic priority, the most relevant military commitments and interventions of the Portuguese Armed Forces were in international crisis management missions, peacekeeping operations and humanitarian assistance missions, in the scope of the various international security organisations of which Portugal is a member: NATO, EU and UN.

Both these new missions and the democratisation underway required a new type of Armed Forces, which led Democracy to launch a reform of the military instrument in order to adapt it to the new reality, defining what may be called a democratic model for its Armed Forces. But what are then the main lines of the reform of the military instrument introduced by Democracy and which characterise the democratic model of the Armed Forces? They are essentially four.

In the first place, the new model of military service, that is, the end of conscription and the training of professional Armed Forces, as well as the inclusion of women in the Armed Forces. The Republic had introduced, in 1911, the model of compulsory and universal military service. Democracy has instituted professional Armed Forces. It was a long and winding path, an inflamed controversy and a hard decision. It followed, however, the trend in other democracies of the NATO and EU allied countries. Since 1991, length of service had been reduced to four months. The 1997 revision removed mandatory service from the constitution, an essential condition for changing the law. Finally, in 1999, the new Law of Military Service ended conscription and established that "in peacetime", the performance of military service relied on volunteering. A transitory period of four years was established for the adjustment of the branches of the Armed Forces, the Army in particular, and from 2004 the professionalisation was complete.

The end of conscription and the adoption of a professional model of armed forces had two sorts of reasons in its origin: one of a technical nature and another of a political nature. The technical reasons had to do with the need for professionalization and technical expertise required by the weapons systems operated by the modern Armed Forces, which entail substantial training time and costs incompatible with a brief passage through the ranks. The political reasons concerned precisely the new type of missions of the Armed Forces, which were not aimed at the traditional defence of the territory but whose purpose, on the contrary, was to ensure international security. Now, if it was seen as legitimate the citizen's obligation to defend the homeland, the same can be said of ensuring international security a long distance away from his own territory. These were the reasons, together with the international trend followed by counterpart forces of the allied countries, which dictated the end of compulsory military service and to the professionalization of the Armed Forces.

This change in the model of military service was accompanied by a significant reduction in the number of military personnel. The nature of the new missions required smaller but more specialised Armed Forces, less geared towards territorial occupation and more expeditionary. Of the around 300,000 men who could be accounted for by the end of the authoritarian regime, during the years of the decolonisation wars, the Portuguese Armed Forced had been reduced to around 80,000 by the early nineties, of which around 46,000 from conscription. In 2010, in a consolidated phase of professionalization, the military personnel neared the 41,000 (22,000 in the Army, 11,000 in the Navy and 8,000 in Air Force).[33]

In the second place, the modernisation of armament and military equipment, with the introduction of planning and financing of modernisation through Military Programming Laws. Historically, the military equipment programmes in Portugal did not obey to long-term planning, let alone to programmed financing. During the last two centuries, the military equipment modernisation programmes were usually haphazard and dictated by the needs of the strategic circumstances and the State's financial willingness. Now, seeing that the processes of military modernisation are slow and require long deadlines, what happened was that, when the moment came for a decision concerning the engagement of the military instrument, the Portuguese Armed Forces did not meet the required conditions. Such was the case during the First World War, in 1916, with belligerence and the intervention in the European theatre, and such was again the case during the Second World War, in 1939, when the state of the Armed Forces was not unconnected to the declaration of neutrality.

Democracy broke up with this cycle and introduced long-term planning and programmed financing in the process of modernisation of the Armed Forces, with the approval of the Military Programming Laws. The first framework law of the Military Programming Laws was approved as early as in 1985 and altered in 1993, and the second framework law was approved in 1998. From that point onwards, the modernisation of the Armed Forces was carried out under these Military Programming Laws, the first, from 1987, revised regularly every three years.[34] It is a fact that, not only the approval of the Laws was constrained by budget restrictions, its execution was on occasion delayed or hampered by mechanisms provided for in the Budget Laws, such as the withholding of funds. But that did not threaten the principle of the Military Programming Laws that represented the fundamental instrument of modernisation of the Armed Forces in a democratic regime.

In the third place, civil-military relations which, in democracy, are characterised by the subordination of the military to the legitimate – that is, democratically elected – political power. Unlike with other democratisation processes, the Portuguese Armed Forces had a key role in the overthrow of the authoritarian regimes, in the transition to democracy and even in democratic consolidation itself. They remained therefore inextricably linked to the foundation of Democracy in Portugal.

Once the imperial cycle was closed, with the decolonisation and given the role of the Armed Forces in the process of democratisation, in 1974 a new cycle of intervention of the military in politics was launched. In a first instance, in a direct fashion, during the revolutionary period between 1974 and 1976. Afterwards, in an indirect but nonetheless effective manner. The 1976 Constitution sprang from a Pact between the MFA (Armed Forces Movement) and the political parties, and enshrined the continuity of military presence in political life through the institutionalisation of the Council of the Revolution. Created by the 25th of April military, it possessed revolutionary entitlement, but lacked democratic legitimacy. However, its powers were broad: first, at the military level, it oversaw defence policy and the Armed Forces; secondly, at the political level, insofar as it held constitutional powers, it oversaw Democracy itself. The Council of the

Revolution and the presence of the military in politics lasted until 1982, the year marking the completion of democratic consolidation and the beginning of a new cycle. A cycle that would span over three decades, marked by the reverse process of the subordination of the Armed Forces to the political power, or, to put it differently, of democratic control over the Armed Forces. This cycle comprehended three crucial moments. The first and most important, as early as in 1982, was the first revision of the Constitution and the approval of the National Defence and Armed Forces Law: the extinction of the Council of the Revolution, the return of the military to their quarters and the creation of the Ministry of National Defence (MDN). The second, in 1993, with the new Organic Basic Laws of the MDN, the Armed Forces General Staff and the branches of the Armed Forces, which transferred responsibilities and powers from the Armed Forces to the MDN, reinforcing political control over the military institution. The third, in 2009, with the approval of a new National Defence Law and, now separately, a new Organic Basic Law of the Organisation of the Armed Forces. It adapted the defence law to the new strategic environment and reinforced the capacity for political leadership of the MDN and the operational command of the Armed Forces Joint Chief of Staff.[35] In the following years, the reforms undertaken followed the same lines.

This was a major change for Democracy – it broke the cycle, dating back from 1820, of intervention by the military in political life, and nowadays constitutes an essential feature of the democratic model of the Armed Forces. "The key reform taken on by the Portuguese Armed Forced is the change in mentality of the military hierarchy, which see themselves as an instrument for action of political power with no legitimacy to challenge it".[36] In Portugal, the subordination of the military institution to legitimate and democratically elected political power is ingrained in the "institutional culture" of the Armed Forces, and is a decisive factor for the quality of Democracy.

After the revolution of 1974 had put an end to the military missions in the colonial theatre, and the role of the military in the democratic regime had ceased in 1982, the Armed Forces went through an identity crisis which was only dispelled in the late eighties, with the beginning of the technical-military cooperation with the PALOP, but especially from the second half of the nineties, with the participation in international missions and peacekeeping operations. And that is precisely the fourth major change, a feature of the democratic model of the Armed Forces.

In the fourth place, then, the new type of missions of the Armed Forces. Since the end of the Cold War, the strategic environment had greatly changed the nature of threats and the type of conflicts. As a consequence, so did the nature of the missions of the Armed Forces change. Constitutionally, the Portuguese Armed Forces are still responsible for the military defence of the Republic. But, in fact, the traditional missions of territorial defence have all but disappeared, while the new missions of ensuring international security are on the rise. Portugal has not been the target of a violation of its mainland territory since the French invasions in the early 19th century, and yet, under a variety of shapes, the ghost of such a threat lingered until the end of the Cold War. In the new post-Cold War strategic environment, Portugal initiated its participation in this new type of missions, which quickly became top priority for the Armed Forces, for defence policy and for the country's foreign policy.

At the start of the process, from the early nineties to 1995, the Portuguese participation in these missions engaged annually around 500 soldiers (average of 475), but between 1999 and 2009, the numbers nearly doubled to an average of 1,000 soldiers, at a minimum of 600 and a maximum of 1,300.[37] Throughout the entire decade, between 1991 and 2000, the average annual expenditure with the Portuguese presence in missions of humanitarian relief and peacekeeping operations neared 50 million euros. In the following decade, keeping up with the reinforcement of the Portuguese Armed Forces' participation in international missions, that average annual expenditure climbed to around 60 million euros.[38]

Portugal became an international security provider and the international missions became a decisive factor, not only for the internationalisation and modernisation of the Armed Forces, but also for the State's external credibility.

Now, this must be why the participation of the Portuguese Armed Forces in international missions were held, in 2010 survey, in such high esteem by Portuguese public opinion (76.8%).[39] Moreover, it was this participation in international peacekeeping missions and the resounding support it was granted by public opinion that contributed to the positive view and the high level of confidence of the Portuguese in their Armed Forces: they are the most trusted national institution, standing at 6.98 in a scale of 0 to 10. (Ahead of the Police (6.46), the Church (6.25), the Media (5.86), the President of the Republic (5.77) and the Courts (5.63). At the bottom of the trust table stand the Unions (5.19) the Parliament (5.14), the government (4.57) and, finally, the political parties (3.35)).[40]

UN peace missions are not a post-Cold War phenomenon. On the contrary, there was a first generation of peacekeeping operations dating back to the post-Second World War period, and the Portuguese

Armed Forces registered their first participation in peacekeeping operations in Lebanon, only three years after Portugal's accession to the UN, in 1958: it was the UNOGIL (United Nations Observation Group in Lebanon) and had the participation of six Portuguese observers. During this first generation of peace missions, between the sixties and the eighties, the UN developed 11 operations worldwide: Congo, Iran, Iraq, New Guinea, Yemen, Cyprus, Dominican Republic, India, Sinai, Golan Heights, Lebanon, Afghanistan e Pakistan.

Portugal, however, was not involved in either of them. The reason was simple enough: the diplomatic friction between Portugal and the UN concerning the colonial question, and, at the military level, the engagement of the Armed Forces in the African war effort, made such an involvement impracticable. During the process of transition and consolidation of democracy, the atmosphere was not yet favourable. With the end of the Cold War and the proliferation of a new type of conflict, on one hand, and the emergence of a more intervening role by the UN in the framework of international security, on the other, a second generation of peace operations arises. It is at this stage, already going through a period of democratic normalisation, that Portugal not only begins to take an active part in this second-generation peace missions, but also makes this participation one of the fundamental axes of its foreign policy and one of the factors of the modernisation and re-establishment of legitimacy of its Armed Forces.

The first of these missions took place in 1989 in Namibia: the UNTAG (United Nations Transition Assistance Group in Namibia), which was carried out between April of 1989 and March of 1990, and relied on the presence of 24 Portuguese military and civil observers.

From that point and until 2010, the Portuguese participation in major peace operations can be discriminated according to three distinct phases and five different theatres: Africa, the Balkans, Timor, Middle East and Central Asia.

The first phase took place during the first half of the nineties and it was marked by UN-led missions directed at the African theatre: in Angola, Mozambique and Western Sahara.

Initially, the Portuguese intervention in Angola's peace process was carried out in the scope of the troika (United States–Russia–Portugal), during the negotiating phase of the process, between 1989 and 1991. From that date and in line with the provisions of the Bicesse Accords, signed in May of 1991, Portugal participated with a mission of international observers in the Angolan elections. This mission was composed by 12 military experts, 6 experts on electoral matters, 5 air traffic controllers and 5 logistic support elements, in a total of 28 observers. In parallel, a military mission was set up led by a three-star general including around 50 soldiers, with the goal of providing training to the Angolan unified Armed Forces. The resumption of hostilities after the elections, in September of 1992, made it necessary to restart the whole peace process and to renegotiate the agreement between the parties, which would be attained through the Lusaka Protocol, in November of 1994. In order to enforce the protocol, and after the UNAVEM II, the UN, in accordance with the Security Council resolution 976 of 1995, set up a new operation, UNAVEM III. Portugal and the Portuguese Armed Forces, under the banner of the UN, take part in UNAVEM III with a substantial military force: 10 officers for the force's staff, including its command; Signal Company No. 5, with 110 elements, Logistics Company No. 6 with 205 elements, four military observers and 20 elements from the National Republican Guard (GNR), in a total of 349 men, within a force of around 6,900 military deployed by 38 countries. This mission, which went on between 1995 and 1997, was replaced by a new UN operation, in accordance with Security Council resolution 1118 of 1997: MONUA, smaller-sized and undertaken between 1997 and 1999. Once again, Portugal was present, on this occasion with a Sanitary Corps with 63 elements, the Signal Company, now with 40 elements, the Logistics Company, 100, and a further two military observers and six GNR elements, including the force's own chief of staff. The operation ended in February of 1999, with the collapse of the peace process.[41]

The second major Portuguese intervention, still in the African theatre, was the peace process in Mozambique, between 1992 and 1994. After the Rome Peace Accords, in October of 1992, Portugal at once took part in the management of the political and military process and in the peace-maintenance force under the command of the UN: the ONUMOZ, implemented by Security Council resolution 797 of 1992. Regarding the first, Portugal deployed a military mission comprising 100 soldiers, led by a two-star general, with a two-fold purpose: to supervise the cease-fire and, above all, to provide training support to the unified Armed Forces of Mozambique. Regarding the ONUMUZ, Portugal participated with elements for the command, military observers and Signal Battalion No. 4, with a total number of soldiers that reached a maximum of 480 soldiers, within a force of around 6,800 soldiers deployed by 40 countries. During the electoral period, Portugal further reinforced its presence with Air Force soldiers and Public Security Police personnel. Brought to an end by the election of the new political bodies – Parliament and Government – and Mozambique's democratic normalisation, this was undoubtedly one of UN's most successful operations.[42]

Also in the scope of the UN, and still in the African theatre, but perhaps lacking the political and military magnitude of the previous ones, Portugal also participated, from 1996, in a third operation: the MINURSO. Instituted by Security Council resolution 690 of 1991, the MINURSO was carried out in Western Sahara and its goals were to provide support and assist in the organisation of the self-determination referendum to determine the status of the territory. Portugal contributed with six officers and, between April of 1996 and June of 1997, the command of the operation was assumed by Portuguese general officers.

The second phase unfolded during the second half of the nineties. Its missions were performed under the banner of NATO and the European Union, and centred on the European theatre: in the Balkans, first in Bosnia and later in Kosovo. Now, if the first three Portuguese interventions had all taken place in the African theatre, the traditional space of strategic interest for Portugal, the same was not the case with the following one. The fourth major Portuguese participation in peace operations dealt with the crisis in the Balkans, from 1996, and marked the opening of a new phase echoing the Europeanist evolution of democratic Portugal.

In December of 1992, faced with the escalation of the crisis in the former Yugoslavia, the Atlantic Council decided for NATO's participation in peacekeeping operations under the *aegis* of the UN. Between 1992 and 1995, the Alliance gave its support to a number of operations intended to implement or to enforce UN resolutions: Security Council resolution 816 of 1992 on the no-fly zone south of Bosnia-Herzegovina (Operation Deny Flight); resolution 836 of 1992 on the protection of Muslim safe areas (the enclaves of Sarajevo and Gorazde); and naval and air operations, whether surveillance and blockade of arms shipment in the Adriatic or targeted air raids at the request of the FORPRONU (operations Sharp Fence, Sharp Vigilance, Maritime Guard and Sharp Guard). When, by the end of 1995, the situation in Bosnia-Herzegovina drastically deteriorated, the Atlantic Alliance, following the Dayton Agreements and mandated by Security Council resolution 1031 of 1995, launched a large-scale operation tending towards peace enforcement and which constituted, at bottom, the first major military operation of NATO. It was called Operation Joint Endeavour, whose mission was assigned to an intervention force, the IFOR. In 1997, the IFOR was replaced by another force, smaller in its numbers but identically organised and with the same objectives. It was instituted by resolution 1088 of 1996. The last phase of the conflict concerned the Kosovo question. In accordance with UN resolution 1244 of 1999, and with the goal of democracy promotion and the creation of a multi-ethnic community in the territory, NATO is assigned the leadership of the process leading to the formation of KFOR.

The complex process in the former Yugoslavia, however, involved from the outset other international actors: the EU, the WEU and the OSCE. Portugal was consistently present and highly engaged in all the activity of the international organisations in the Balkans. Besides its participation in the EU's missions to monitor the implementation of the Brioni Agreements, since 1991, Portugal integrated, between 1992 and 1995, in the framework of NATO and the WEU, air space restriction and naval blockade operations in the Adriatic, with two Air Force airplanes (P3-Orion to assist the blockade and C-130 for logistic support) and a Navy ship on a permanent basis. Rotationally, a frigate and a corvette. And, from October and December of 1993, one submarine. During the Portuguese presidency of the WEU, in the first semester of 1992, Portugal sent 30 observers, and in February of 1993 and until 1994, a medical team to Bihac, to provide support to the French contingent. Its most significant participation, however, will be developed in the framework of NATO, from 1996.

Portugal participated with IFOR in the implementation of the Dayton Agreements, between January and December of 1996, with an infantry battalion from the Independent Airborne Brigade and the Support Service Detachment in a total of 919 soldiers, incorporated in the Italian brigade in action in the French sector in the Sarajevo area and in the Gorazde pocket, in Rogatica. Since January of 1997, Portugal maintained with the SFOR a reinforced and airborne infantry company with a total of 320 soldiers.[43]

In addition to these forces mainly detached from the Army, the Air Force sent a C-212 and a C-130 to give support to naval forces, and took part on Operation Deny Flight with advanced air traffic controllers and military observers, in a total of 60 elements.

Operation Eagle Eye was carried out from October of 1998 to April of 1999, and involved aerial surveillance of the territory of the Federal Republic of Yugoslavia with the main purpose of verifying compliance with the UN resolutions on Kosovo. Portugal participated with three F-16 and around 60 soldiers, between pilots and support personnel, headquartered at the NATO base of Aviano, in northern Italy. In May of 199, Operation Allied Force was launched in Yugoslavia, with Portugal maintaining the same degree of engagement in air patrol. Finally, following UN resolution 1244, passed in June of 1999, KFOR was set up under NATO's leadership, and Operation Joint Guard was launched. Incorporated in a force of 50,000 soldiers from 30 countries, Portugal participated with a contingent of 323 soldiers, who began to arrive in the theatre of operations in August of 1999. The Portuguese contingent, which had reached its full number in October, was distributed across the headquarters and staffs, had liaison elements with KFOR and the Italian brigade, and integrated, in addition to a special operations detachment – the Agrupamento Bravo, headquartered in Kilna – the support services and a team of advanced air traffic controllers. In February of 2000, the force was slightly downsized, keeping from then on a total of 295 soldiers – Task Force Pegasus.[44]

The Portuguese presence in the theatre of operations of the Balkans had major political significance and marked a turning point in the tradition of the State's foreign policy. It was in line with Portugal's full assumption of its statute as a European country and with the most striking phase of Europeanisation, not only of foreign policy, but also of defence policy in Portugal. From the point of view of the international presence of the Armed Forces, it broke with an already long military tradition and also with a long lineage of strategic thought, and it was the first time, since the Great War of 1915–1918, that the Portuguese troops intervened in military operations in the European theatre.[45]

The third phase developed from 2000. Never had the Portuguese Armed Forces been more engaged in this type of operations. The missions multiplied, there was a diversification in the theatres of operations and there were missions being carried out at the same time in the scope of NATO, of the EU and of the UN. Portugal did not withdraw either from the African or the European theatre, but became engaged in new theatres. Some of them, such as East Timor, corresponded to the historical interest ensuing from Portugal's post-colonial relations, while others were simply a matter of siding with a common concern to provide international security, such as Afghanistan or Lebanon.

The fifth great Portuguese intervention in UN peace operations once more widened the area of intervention of the Portuguese Armed Forces, and concerned the process in East Timor. In June of 1999, through Security Council resolution 1246, UNAMET (United Nations Mission to East Timor) was set up with the purpose of organising and supervising the referendum on Timor's political future, indeed preserving the bilateral agreement of the 5th of May between Portugal and Indonesia, as well as the agreements between the two governments and the United Nations. Following a UN resolution, the Portuguese government decided to create a mission of observers to accompany the process of popular consultation in East Timor. Formed by 50 elements, the Portuguese missions left Lisbon on the 13th of June and extended its functions until the 10th of September.

Following the violent events that took place after the referendum, whose outcome clearly expressed their vote in favour of the independence of the territory, the UN, through Security Council resolution 1264, of the 13th of September of 1999, created the Multinational Force for East Timor (INTERFET), with an initial contingent of 4,600 soldiers, the majority of them Australian (3,534) and under the command of Australia. With no diplomatic insistence, given the delicate political situation, Portugal made available to the force a frigate from its Armada. It never officially integrated INTERFET.

In the course of the complex process that led to Timor's independence, the UN, through its resolution 1272, adopted on the 22th of October of 1999, established and mandated a mission with leading the transition process and the administration of the territory until Timor's independence on the 20th of May of 2002: the UNTAET (United Nations Transnational Administration for East-Timor). In the UNTAET, Portugal consistently maintained a military contingent of around 1000 soldiers. Towards the end of the mission, in early 2002, the full number of men was reached, distributed over a joint infantry battalion (PORBAT) with 709 soldiers; a Marine company of 155 soldiers; a detachment of helicopters with 28 soldiers and four aircrafts; in addition, 47 soldiers integrated in the force headquarters and in the command of the central sector, which stood under the responsibility of the Portuguese battalion.

Even though it was not properly a new major operation, EUFOR-ALTHEA became the successor to operation SFOR in December of 2004, with the transfer of responsibility from NATO to the European Union. Its goal was to maintain security in Bosnia and Herzegovina until long-term stability was guaranteed. Portugal was as much engaged in EUFOR-ALTHEA as it had been in SFOR, with a battalion-level unit, to which were later added two Liaison Observation Teams (LOT). Portugal ended its engagement in this operation in late 2011.

The sixth major Portuguese intervention in peace operations took place in Central Asia and, once again, in a theatre totally unrelated to the traditional areas of the national strategic interest. In 2001, in the context of the so-called "war on terror", Security Council 1386 provided for the creation of the ISAF (International Security Assistance Force). This was a NATO-led security mission intended to intervene in Afghanistan and composed by some twenty countries, both members of the Alliance and non-members. In 2002, ISAF's first mission was insuring security in Kabul, but from 2003 it expanded its intervention throughout Afghanistan. Between 2006 and 2009, it had its more intensive phase in terms of combat and later evolved to training functions, namely the training of the Afghan Armed Forces.

Portugal integrated the ISAF in 2002 with a rotating company-level force composed by Commandos or Paratroopers and with a contingent oscillating between 160 and 200 soldiers. In the toughest phase of the combat, the Portuguese forces led with no "caveats" operations in various zones: besides Kabul, Kandahar, Farah and Herat. When the mission progressed to the training of Afghan forces and the so-called Orientation Mentor Liaison Teams (OMLT) were created, Portugal also participated in those operations, integrating the mission until its completion.

The seventh intervention of the Portuguese Armed Forces in peace operations once again took place in a theatre of no historical affinity or geographic proximity to the areas of national strategic interest: the Middle East. Yet again, the logic of international security provision prevailed. Since 1978, under Security Council resolutions 425 and 426, the UN had been conducting a peace operation in Lebanon with the purpose of ensuring the withdrawal of Israeli forces from South Lebanon and assisting the Lebanese State in restoring authority over its territory: the UNIFIL (United Nations Interim Force in Lebanon). The UNIFIL's mandate had already been the object of two updates, in 1982 and in 2000, but in 2006, following another crisis between Israel and the Hezbollah, the Security Council decided to reinforce the mandate and reorganise UNIFIL's force, which implied a significant enhancement in its contingent. In this scope, and reaffirming the priority given to participation in international missions, Portugal decided to integrate the UNIFIL. On this occasion, with a company-level contingent – an engineer unit composed of 141 soldiers. It arrived in Lebanon in late 2006 and finished its mission in June of 2012.

Finally, in the scope of a global counter-piracy action in the Horn of Africa, the European Union launched, in December of 2008, Operation ATLANTA, with the purpose of ensuring maritime security in the west coast of the Indian Ocean. Portugal participated in this operation from the outset with EUNAVFOR Staff officers. From April of 2010, it contributed with a P3 patrol airplane and, on three occasions, with naval means, having twice taken on the command of the operation.

In parallel to these larger-scale operations, Portugal had, in the course of this period and until 2010, a part in many other international missions, deploying either specific units or individual soldiers for different missions.[46]

The Portuguese presence in international mission had the greatest significance and profound consequences not only for the Armed Forces, but also for Portugal's defence policy and even foreign policy. In the first place, the participation of the Portuguese Armed Forces in missions of humanitarian assistance and crisis management and peace operations had a huge impact at the level of defence policy. Participation in that new type of missions represented a crucial factor for the technical and professional modernisation of the military instrument, as well as for the renewal of legitimacy and prestige of the Armed Forces in Portuguese society.[47]

Finally, and more important still, it meant for Portugal the relinquishment of its traditional position as liquid consumer of international security and the adoption of an active position as international security provider, with all its implications in terms of international prestige for the country and external credibility for the State.

Notes

1 Between Liberal Revolution and Civil Wars

1 MACEDO, Jorge, *História Diplomática Portuguesa – constantes e linhas de força*, Lisboa, Instituto Defesa Nacional, 1987.

2 BONIFÁCIO, Fátima, *Seis Estudos sobre o Liberalismo português*, Lisboa, Editorial Estampa, 1991, pp. 327–328; SARDICA, José Miguel, "A Revolução Liberal e a política de defesa e segurança, 1820–1851", in João Marques de ALMEIDA; Rui RAMOS, *Revoluções, Política Externa e Política de Defesa em Portugal, Séc. XIX–XX*, Lisboa, Edições Cosmos/Instituto Defesa Nacional, 2008, p. 30.

3 BONIFÁCIO, Fátima, "Relações externas e Revolução Liberal, Séc. XVIII-XIX", in João Marques de ALMEIDA; Rui RAMOS, *Revoluções, Política Externa e Política de Defesa em Portugal, Séc. XIX–XX*, Lisboa, Edições Cosmos/Instituto Defesa Nacional, 2008, p. 22.

4 VENTURA, António, *As Guerras Liberais, 1820–1834*, Lisboa, Academia Portuguesa de História, 2008, p. 18.

5 BRANDÃO, Raul, *El-Rei Junot*, Lisboa, IN-CM, 1982.

6 SARDICA, José Miguel, "A Revolução Liberal e a política de defesa e segurança, 1820–1851", 2008, pp. 26–3l; BEBIANO, Rui, "Organização e Papel do Exército", in José MATTOSO, *História de Portugal*, Lisboa, Círculo de Leitores, 1993, pp. 253–255; MARQUES, Fernando Pereira, *Exército, mudança e modernização na primeira metade do século XIX*, Lisboa, Edições Cosmos/Instituto da Defesa Nacional, 1999, pp. 25–35.

7 SELVAGEM, Carlos, *Portugal Militar*, Lisboa, IN-CM, 2006, p. 577.

8 MARQUES, Fernando Pereira, *Exército, mudança e modernização na primeira metade do século XIX*, 1999, p. 46.

9 SELVAGEM, Carlos, *Portugal Militar*, 2006, p. 577; BEBIANO, Rui, "Organização e Papel do Exércio", 1993, pp. 255–257; MARQUES, Fernando Pereira, *Exército, mudança e modernização na primeira metade do século XIX*, 1999, pp. 46–48.

10 TELO, António José, *História da marinha portuguesa. Homens, doutrinas e organização, 1824–1974*, tomo 1, Lisboa, AM, 1999, pp. 28–29

11 MARQUES, Fernando Pereira, *Exército, mudança e modernização na primeira metade do século XIX*, 1999, p. 49.

12 For a global analysis of military interventions in politics between the 1820 revolution and the Regeneration, see VALENTE, Vasco Pulido, *Os Militares e a Politica (1820–1856)*, Lisboa, IN-CM, 1997. Specifically on the 1820 revolution, pp. 59–64.

13 VALENTE, Vasco Pulido, *Os Militares e a Politica (1820–1856)*, 1997, pp. 64–69.

14 SELVAGEM, Carlos, *Portugal Militar*, 2006, p. 577; BEBIANO, Rui, "Organização e Papel do Exército", 1993, p. 258; MARQUES, Fernando Pereira, *Exército, mudança e modernização na primeira metade do século XIX*, 1999, pp. 48–49 and 130.

15 TELO, António José, *História da Marinha Portuguesa. Homens, doutrinas e organização, 1824–1974*, 1999, pp. 30 and 23–24.

16 MARQUES, Fernando Pereira, *Exército, mudança e modernização na primeira metade do século XIX*, 1999, p. 131.

17 SOUSA, Bernardo Vasconcelos and; RAMOS, Rui; MONTEIRO, Nuno Gonçalo, *História de Portugal*, Lisboa, Esfera dos Livros, 2009, p. 478.

18 VENTURA, António, *As Guerras Liberais, 1820–1834*, 2008, pp. 25–31.

19 SELVAGEM, Carlos, *Portugal Militar*, 2006, p. 577.

20 TELO, António José, *História da marinha portuguesa. Homens, doutrinas e organização, 1824–1974*, 1999, pp. 30 and 42–43.

21 VENTURA, António, *As Guerras Liberais, 1820–1834*, 2008, p. 40; SELVAGEM, Carlos, *Portugal Militar*, 2006, p. 549; VALENTE, Vasco Pulido, *Os Militares e a Politica (1820–1856)*, 1997, pp. 97–105.

22 SOUSA, Bernardo Vasconcelos and; RAMOS, Rui; MONTEIRO, Nuno Gonçalo, *História de Portugal*, 2009, p. 485.

23 VENTURA, António, *As Guerras Liberais, 1820–1834*, 2008, p. 50.

24 VALENTE, Vasco Pulido, *Os Militares e a Politica (1820–1856)*, 1997, pp. 105–116.

25 SELVAGEM, Carlos, *Portugal Militar*, 2006, p. 577.

26 BEBIANO, Rui, "Organização e Papel do Exército", 1993, p. 26; MARQUES, Fernando Pereira, *Exército, mudança e modernização na primeira metade do século XIX*, 1999, p. 132; SELVAGEM, Carlos, *Portugal Militar*, 2006, pp. 547–562.

27 TELO, António José, *História da marinha portuguesa. Homens, doutrinas e organização, 1824–1974*, 1999, p. 55.

28 VENTURA, António, *As Guerras Liberais, 1820–1834*, 2008, pp. 105–121.

29 TELO, António José, *História da marinha portuguesa. Homens, doutrinas e organização, 1824–1974*, 1999, pp. 78–79.

30 VALENTE, Vasco Pulido, *Os Militares e a Politica (1820–1856)*, 1997, pp. 116–117.

31 *Ibidem*, pp. 117–118.

32 *Ibidem*, pp. 118–127.

33 MARQUES, Fernando Pereira, *Exército, mudança e modernização na primeira metade do século XIX*, 1999, pp. 132–133; SELVAGEM, Carlos, *Portugal Militar*, 2006, p. 578.

34 SELVAGEM, Carlos, *Portugal Militar*, 2006, pp. 578–579; BEBIANO, Rui, "Organização e Papel do Exército", 1993, p. 260; MARQUES,

Fernando Pereira, *Exército, mudança e modernização na primeira metade do século XIX*, 1999, pp. 50–51.

35 SOUSA, Bernardo Vasconcelos and; RAMOS, Rui; MONTEIRO, Nuno Gonçalo, *História de Portugal*, 2009, p. 504; BONIFÁCIO, Maria de Fátima, *Um Homem Singular. Biografia Política de Rodrigo da Fonseca Magalhães, 1787–1858*, Lisboa, D. Quixote, 2013.

36 SELVAGEM, Carlos, *Portugal Militar*, 2006, p. 573; VENTURA, António, *Da Revolução de 1820 ao fim das Guerras Civis*, in Manuel Themudo BARATA; Nuno Severiano TEIXEIRA, *Nova História Militar de Portugal*, vol. III, Lisboa, Círculo de Leitores, 2004, pp. 225–227; VALENTE, Vasco Pulido, *Os Militares e a Politica (1820–1856)*, 1997, pp. 127–132.

37 SOUSA, Bernardo Vasconcelos; RAMOS, Rui and; MONTEIRO, Nuno Gonçalo, *História de Portugal*, 2009, p. 516.

38 BONIFÁCIO, Maria de Fátima, *História da Guerra Civil da Patuleia, 1846–1847*, Lisboa, Editorial Estampa, 1993; SELVAGEM, Carlos, *Portugal Militar*, 2006, pp. 573–576; VENTURA, António, *Da Revolução de 1820 ao fim das Guerras Civis*, 2004, pp. 227–229.

39 SELVAGEM, Carlos, *Portugal Militar*, 2006, p. 577; BEBIANO, Rui, "Organização e Papel do Exército", 1993, pp. 261–262; MARQUES, Fernando Pereira, *Exército, mudança e modernização na primeira metade do século XIX*, 1999, pp. 51–136.

40 SARDICA, José Miguel, "Os Militares e a Política entre o Ultimato e a República", in Fernando MARTINS (ed.), *Diplomacia & Guerra*, Lisboa, Edições Colibri, 2001, pp. 11–13; VALENTE, Vasco Pulido, *Os Militares e a Politica (1820–1856)*, 1997.

41 MARTINS, Oliveira, *Portugal Contemporâneo*, vol. 2, Lisboa, Guimarães Editora, 1986, p. 121.

2 Building the African Empire

1 ALEXANDRE, Valentim (coord.), "O Império africano (séculos XIX–XX) – As linhas gerais", in *Império Africano – séculos XIX–XX*, Lisboa, Edições Colibri, 2000, pp. 12–15; COSTA, João Paulo; RODRIGUES, José Damião; OLIVEIRA, Pedro Aires (coord.), *História da Expansão e do Império Português*, Lisboa, Esfera dos Livros, 2014, pp. 349–366.

2 ALEXANDRE, Valentim, *Origens do Colonialismo Português Moderno*, Lisboa, Sá da Costa, 1979, pp. 159–161 and 162–163.

3 Pedro Tavares de Almeida – ofício n.º 24 611, AHU, Conselho Ultramarino, in ALEXANDRE, Valentim, *Origens do Colonialismo Português Moderno*, 1979, p. 58.

4 COSTA, João Paulo; RODRIGUES, José Damião; OLIVEIRA, Pedro Aires (coord.), *História da Expansão e do Império Português*, 2014, pp. 367–376; ALEXANDRE, Valentim (coord.), "O Império africano (séculos XIX-XX) – As linhas gerais", 2000, pp. 15–17; ALEXANDRE, Valentim, "Nação e Império", in Francisco BETHENCOURT; Kirti

CHAUDHURI, *História da Expansão Portuguesa*, vol. IV, Lisboa, Círculo de Leitores, 1998, pp. 90–104.

5 MIÈGE, Jean-Louis, *Expansion Européenne et décolonisation de 1870 à nous jours*, Paris, Puf, 1973, pp. 143–144.

6 GERARD, C. L.; LESOURD, J. A., *La nouvelle histoire economique*, vol. 1, Paris, Librairie Armand Colin, 1976, p. 104.

7 MIÈGE, Jean-Louis, *Expansion Européenne et décolonisation de 1870 à nous jours*, 1973, pp. 148–149.

8 To this respect, see GERARD, C. L.; LESOURD, J. A., *La nouvelle histoire economique*, 1976, pp. 33 and ff.; BOUVIER, Jean, *Initiation au Vocabulaire et aux Mécanismes Économiques Contemporains,* Paris, SEDES, 1977, pp. 27–32 ; and especially NIVEAU, Maurice, *Histoire des Faits Économiques Contemporains,* Paris, Presses universitaires de France, 1966, pp. 143–189, particularly pp. 179–181.

9 To this respect, see RENOUVIN, Pierre, *Histoire des Relations Internationales*, Paris, Hachete, 1994, pp. 34–45; e MILZA, Pierre, *Les Relations Internationales de 1871 à 1914,* Paris, Librairie Armand Colin, 1968, pp. 57–65.

10 DUROSELLE, Jean-Baptiste, *L'Europe de 1815 à nos Jours,* Paris, Puf, 1964, p. 137.

11 To this respect, see GUIMARÃES, Ângela, "A ideologia colonialista em Portugal no último quartel do século XIX", in *Ler História, n.º* 1, de Janeiro/Abril, 1983, pp. 69–79; and GUIMARÃES, Ângela, *Uma Corrente do Colonialismo Português,* Lisboa, Livros Horizonte, 1984; CAETANO, Marcello, *Tradições, Princípios e Métodos da Colonização Portuguesa,* Lisboa, Agência do Ultramar, 1951; CASTELO, Cláudia, *O Modo Português de Estar no Mundo. O Luso-Tropicalismo e a Ideologia Colonial Portuguesa (1933–1961)*, Lisboa, Edições Afrontamento, 1998. Foreign researchers were also sensitive to this feature of Portuguese colonialismo. See, for instance, HAMMOND, Richard, *Portugal and Africa 1815–1910. A Study in Uneconomic Imperialism,* Stanford-California, 1966; and, drawing attention to the economic value of the Empire, CLARENCE-SMITH, Gervase, *The Third Portuguese Empire,* Manchester, Manchester University Press, 1985.

12 On the Portuguese exploration voyages into the hinterlands of the African continent, see SANTOS, M. Emília Madeira, *Viagens de Exploração Terrestre dos Portugueses em África,* Lisboa, Centro de Estudos de Cartografía Antiga, 1978, pp. 175 and ff.

13 CORDEIRO, Luciano, *Obras de Luciano Cordeiro: Questões coloniais,* vol. 1, Coimbra, 1934, p. 370.

14 RENOUVIN, Pierre, *Histoire des Relations Internationales*, 1994, pp. 46–55.

15 *The Times,* de 15 de Maio de 1884.

16 The colonial race and the partition of Africa elicited, from 1961, a long theoretical debate launched by the work by ROBINSON, Ronald; GALLAGHER, John; DENNY, Alice, *Africa and the Victorians: The*

Official Mind of Imperialism, Londres, Macmillan, 1961, and followed by *Journal of African History,* it extends until the seventies with the theses by BRUNSCHWIG, Henri, *Le Partage de l'Afrique Noire,* Paris, Flammarion, 1971. The Portuguese position in this setting was studied by AXELSON, Eric, *Portugal and the Scrumble for Africa 1875–1891,* Joanesburgo, Witwatersrand University, 1967.

17 Salisbury, in MIÈGE, Jean-Louis, *Expansion Européenne et décolonisation de 1870 à nous jours*, 1973, p. 181.

18 ALEXANDRE, Valentim; Dias, Jill, "O Império Africano 1825–1890", tomo X, in Joel SERRÃO; A. H. Oliveira MARQUES (dir.), *Nova História de Portugal*, Lisboa, Editorial Estampa, 1998, pp. 112–113.

19 SANTOS, M. Emília Madeira, *Viagens de Exploração Terrestre dos Portugueses em África*, 1978.

20 ALEXANDRE, Valentim, "Nação e Império", 1998, pp. 117–118.

21 ALEXANDRE, Valentim, "Nação e Império", 1998, pp. 106–108.

22 Tratado do Zaire em ALMADA, José de, *Tratados Aplicáveis ao Ultramar,* Lisboa, vol. Principia, 1943, p. 7719.

23 TELO, António José, *Lourenço Marques na política externa portuguesa 1875–1900*, Lisboa, Cosmos, 1991.

24 CAETANO, Marcello, *Portugal e a Internacionalização dos Problemas Africanos,* Lisboa, Ática, 1971.

25 COSTA, João Paulo; RODRIGUES, José Damião; OLIVEIRA, Pedro Aires (coord.), *História da Expansão e do Império Português*, 2014, pp. 390–397; ALEXANDRE, Valentim (coord.), "O Império africano (séculos XIX–XX) – As linhas gerais", 2000, pp. 17–18; ALEXANDRE, Valentim, "Nação e Império", 1998, pp. 108–112 and 119–125.

26 For a summary, see ALMADA, José de, *Tratados Aplicáveis ao Ultramar*, 1943, pp. 30 and ff.

27 About the juridic definition of the sphere of influence and its consequences for the international scene, see CAETANO, Marcello, *Portugal e a Internacionalização dos Problemas Africanos, 1971,* p. 129.

28 COSTA, João Paulo; RODRIGUES, José Damião; OLIVEIRA, Pedro Aires (coord.), *História da Expansão e do Império Português*, 2014, pp. 399–403.

29 A summary of this exchange can be found in CHAGAS, Pinheiro, *Um Reinado Trágico,* Lisboa, vol. I, 1908, pp. 92–107; ALEXANDRE, Valentim, "Nação e Império", 1998, pp. 128–130; TEIXEIRA, Nuno Severiano, "Colónias e colonização portuguesa na cena internacional (1885–1930)", in Francisco BETHENCOURT; Kirti CHAUDHURI, *História da Expansão Portuguesa*, vol. IV, Lisboa, Círculo de Leitores, 1998, pp. 502–503.

30 TEIXEIRA, Nuno Severiano, *O Ultimatum Inglês – política externa e polítca interna no Portugal de 1890*, Lisboa, Alfa, 1990, pp. 60–61.

31 These diplomatic notes can be found in VILHENA, Júlio de, *Antes da República*, vol. I, Lisboa, França e Arménio, 1916, pp. 182 and 183.

32 For the Treaty's text (as well as for the proposed ammendments), see

ALMADA, José de, *Tratados Aplicáveis ao Ultramar*, 1943, pp. 342–351.

33 To this respect, check the diplomatic correspondence between London and Lisbon in ALMADA, José de, *Tratados Aplicáveis ao Ultramar*, 1943, pp. 353–364 and ff.

34 For the Treaty's published texto, see ALMADA, José de, *Tratados Aplicáveis ao Ultramar*, 1943, pp. 389–400.

35 TEIXEIRA, Nuno Severiano, *O Ultimatum Inglês – política externa e política interna no Portugal de 1890*, 1990.

36 SELVAGEM, Carlos, *Portugal Militar*, 2006, p. 579; BEBIANO, Rui, "Organização e Papel do Exércio", 1993, p. 262.

37 SELVAGEM, Carlos, *Portugal Militar*, 2006, pp. 579–580; BEBIANO, Rui, "Organização e Papel do Exército", 1993, p. 262.

38 TELO, António José, *História da marinha portuguesa. Homens, doutrinas e organização, 1824–1974*, 1999, pp. 186–201; SELVAGEM, Carlos, *Portugal Militar*, 2006, pp. 581–582.

39 SARDICA, José Miguel, "Os Militares e a Política entre o Ultimato e a República", 2001, pp. 14–18.

40 FERREIRA, José Medeiros, *O comportamento político dos militares – forças armadas e regimes políticos em Portugal no século XX*, Lisboa, Estampa, 1992, pp. 29 and ff.

41 SELVAGEM, Carlos, *Portugal Militar*, 2006, pp. 591–592.

42 COSTA, João Paulo; RODRIGUES, José Damião; OLIVEIRA, Pedro Aires (coord.), *História da Expansão e do Império Português*, 2014, p. 419.

43 CARRILHO, Maria, *Forças Armadas e mudança política em Portugal no séc. XX – Para uma explicação sociológica do papel dos militares*, Lisboa, IN-CM, 1985, p. 151; SARDICA, José Miguel, "Os Militares e a Política entre o Ultimato e a República", 2001, pp. 28–29.

44 According to Pélissier's periodisation: the time of the Centurions (1890–1909) and the time of the clearances (1910–1920), in PÉLISSIER, René, *As Campanhas Coloniais de Portugal*, Lisboa, Estampa, 2004.

45 COSTA, João Paulo; RODRIGUES, José Damião; OLIVEIRA, Pedro Aires (coord.), *História da Expansão e do Império Português*, 2014, p. 416.

46 In a total of 1022, 477 of which metropolitan, in ALEXANDRE, Valentim, "Situações Coloniais: II – O Ponto de Viragem: as campanhas de ocupação (1890–1930)", in Francisco BETHENCOURT; Kirti CHAUDHURI, *História da Expansão Portuguesa*, vol. IV, Lisboa, Círculo de Leitores, 1998, p. 189.

47 ALEXANDRE, Valentim, "Situações Coloniais: II – O Ponto de Viragem: as campanhas de ocupação (1890–1930)", 1998, p. 190.

48 PÉLISSIER, René, *História de Moçambique – formação e oposição, 1854–1918*, 2 vols., Lisboa, Editorial Estampa, 1987; ALEXANDRE, Valentim, "Situações Coloniais: II – O Ponto de Viragem: as campanhas de ocupação (1890–1930)", 1998, pp. 182–193; BESSA, Carlos,

"Moçambique, Ocupação Efectiva Peculiar", in Manuel Themudo BARATA; Nuno Severiano TEIXEIRA, *Nova História Militar de Portugal*, vol. III, Lisboa, Círculo de Leitores, 2004, pp. 287–306; SELVAGEM, Carlos, *Portugal Militar*, 2006, pp. 593–642.

49 There was a total of 250 killed or missing, in ALEXANDRE, Valentim, "Situações Coloniais: II – O Ponto de Viragem: as campanhas de ocupação (1890–1930)", 1998, p. 195.

50 PÉLISSIER, Réne, *História das Campanhas de Angola – Resistência e revoltas (1945–1941)*, 2 vols., Lisboa, Editorial Estampa, 1986; ALEXANDRE, Valentim, "Situações Coloniais: II – O Ponto de Viragem: As campanhas de ocupação (1890–1930)", 1998, pp. 193–202; BESSA, Carlos, "Angola, do feudalismo luso-africano à ocupação", in Manuel Themudo BARATA; Nuno Severiano TEIXEIRA, *Nova História Militar de Portugal*, vol. III, Lisboa, Círculo de Leitores, 2004, pp. 271–286; SELVAGEM, Carlos, *Portugal Militar*, 2006, pp. 643–681.

51 PÉLISSIER, René, *História da Guiné – portugueses e africanos Senegâmbia (1841–1936)*, 2 vols., Lisboa, Editorial Estampa, 1989; ALEXANDRE, Valentim, "Situações Coloniais: II – O Ponto de Viragem: as campanhas de ocupação (1890–1930)", 1998, pp. 202–206; BESSA, Carlos, "Guiné, das feitorias isoladas ao "enclave" unificado", in Manuel Themudo BARATA; Nuno Severiano TEIXEIRA, *Nova História Militar de Portugal*, vol. III, Lisboa, Círculo de Leitores, 2004, pp. 257–270; BESSA, Carlos, "Timor. Do domínio liurai à pacificação portuguesa", in Manuel Themudo BARATA; Nuno Severiano TEIXEIRA, *Nova História Militar de Portugal*, vol. III, Lisboa, Círculo de Leitores, 2004, pp. 323–333; SELVAGEM, Carlos, *Portugal Militar*, 2006, pp. 681–685.

3 The Republic and the Great War

1 LANGHORNE, Richard, "Anglo-German negotiation concerning the future of the Portuguese colonies, 1911–1914", in *The Historical Journal*, vol. XVI, n.º 2, 1973, pp. 361–387; Willequet, Jacques, "Anglo-German rivalry in Belgian and Portuguese Africa", in *Britain and German in Africa, imperial rivalry and colonial rule*, New Haven/London, 1967, pp. 245–273; SMITH, John Vincent, "The Anglo-German negotiation over the Portuguese colonies in Africa, 1911–1914", in *The Historical Journal*, vol. XVII, n.º 3, 1974, pp. 562–629.

2 TEIXEIRA, Nuno Severiano, "Portugal no contexto internacional", in, Francisco BETHENCOURT; Kirti CHAUDHURY (dirs.), *História da Expansão Portuguesa*, vol. IV, Lisboa, Círculo de Leitores, 1998, pp. 494–520.

3 COSTA, Fernando, *Portugal e a Guerra Anglo-Bóer: política externa e opinião pública (1899–1902)*, Lisboa, Cosmos, 1998.

4 TORRE GÓMEZ, Hipólito de la, *El Imperio del Rey. Alfonso XIII, Portugal y los Ingleses (1907–1916)*, Mérida, Editora Regional de Extremadura, Junta de Extremadura/Unión Europea, 2002.

5 SMITH, John Vincent, *As relações políticas luso-britânicas, 1910–1916*, Lisboa, Livros Horizonte, 1975; DEROU, Jean, *Les relations Franco-Portugaises 1910–1926*, Paris, Publications de la Sorbonne, 1986; TELO, António José, *Os Açores e o controlo do Atlântico (1898/1948)*, Porto, Asa, 1993.

6 TORRE GÓMEZ, Hipólito de la, *Conspiração contra Portugal (1910–1912)*, Lisboa, 1978.

7 TEIXEIRA, Nuno Severiano, *O poder e a guerra, 1914–1918. Objectivos nacionais e estratégias políticas na entrada de Portugal na Grande Guerra*, Lisboa, Estampa, 1996.

8 CIDADE, Hernâni, "Portugal na Guerra Mundial: 1914–1918", in Damião PERES (dir.), *História de Portugal*, vol. VII, Barcelos, Portucalense, 1933, pp. 491–522; MARTINS, Ferreira, *Portugal na Grande Guerra*, Lisboa, Ática, 1934–38.

9 TORRE GÓMEZ, Hipólito de la, *Na encruzilhada da Grande Guerra: Portugal-Espanha, 1913–1919*, Lisboa, Estampa, 1980; FERREIRA, José Medeiros, *Um século de problemas. As relações luso-espanholas da União Ibérica à Comunidade Europeia*, Lisboa, Livros Horizonte, 1989, pp. 31–35.

10 VALENTE, Vasco Pulido, "Portugal e a Guerra de 14–18 (1)", in *O Tempo e o Modo*, n.º 33, 1965, pp. 1202–1215; VALENTE, Vasco Pulido, *A "República Velha" (1910–1917)*, Ensaio, Lisboa, Gradiva, 1997; TEIXEIRA, Nuno Severiano, *O poder e a guerra, 1914–1918. Objectivos nacionais e estratégias políticas na entrada de Portugal na Grande Guerra*, 1996; TELO, António José, *O sidonismo e o movimento operário: Luta de classes em Portugal, 1917–1919*, Lisboa, Ulmeiro, 1977, CABRAL, Manuel Villaverde, "A Grande Guerra e o sidonismo (esboço interpretativo)", in *Análise Social*, vol. XV, n.º 58, 1979, pp. 373–392. More recently, TELO, António José; SOUSA, Pedro Marquês de, *O CEp. Os militares sacrificados pela má política*, Lisboa, Fronteira do Caos, 2016.

11 VALENTE, Vasco Pulido, *A "República Velha" (1910–1917)*, 1997; LOPES, Fernando Farelo, *Poder político e caciquismo na 1.ª República Portuguesa*, Lisboa, Estampa, 1994; MARQUES, A. H. de Oliveira (coord.), *Portugal da Monarquia para a República*, vol. XI, in Joel SERRÃO; A. H. de Oliveira MARQUES (dir.), *Nova História de Portugal*, Lisboa, Presença, 1991.

12 VALENTE, Vasco Pulido, "Revoluções: A "República Velha" (ensaio de interpretação política)", in *Análise Social*, vol. XXVII, n.º 115 (1.º), 1992, pp. 7–63; VALENTE, Vasco Pulido, *A "República Velha" (1910–1917)*, 1997; TEIXEIRA, Nuno Severiano, *O poder e a guerra, 1914–1918. Objectivos nacionais e estratégias políticas na entrada de Portugal na Grande Guerra*, 1996; MENESES, Filipe Ribeiro de, *União Sagrada e sidonismo – Portugal na guerra (1916–18)*, Lisboa, Cosmos, 2000; FRAGA, Luís M. Alves, *O fim da ambiguidade: a estratégia nacional portuguesa de 1914 a 1916*, Lisboa, Universitária, 2001.

13 *Palavras claras. Razões da intervenção militar de Portugal na guerra europeia.* Report published in "Diário do Governo" n.º 9, 1.ª série de 17 de Janeiro de 1917, Lisboa, Imprensa Nacional, 1917.

14 TELO, António José, *Decadência e queda da I República Portuguesa*, 1.º vol. , Lisboa, A Regra do Jogo, 1980; VALENTE, Vasco Pulido, *O poder e o povo. A Revolução de 1910*, Lisboa, Moraes, 1982; CARRILHO, Maria, *Forças Armadas e mudança política em Portugal no séc. XX – Para uma explicação sociológica do papel dos militares*, Lisboa, IN-CM, 1985; FERREIRA, José Medeiros, *Portugal na Conferência de Paz. Paris, 1919*, Lisboa, Quetzal, 1992.

15 CARRILHO, Maria, *Forças Armadas e mudança política em Portugal no séc. XX – Para uma explicação sociológica do papel dos militares*, 1985, p. 209.

16 TELO, António José, *Decadência e queda da I República Portuguesa*, 1.º vol. , Lisboa, A Regra do Jogo, 1980, p. 103.

17 TELO, António José, *História da marinha portuguesa. Homens, doutrinas e organização, 1824–1974*, tomo I, Lisboa, AM, 1999, p. 232.

18 MARTINS, Ferreira, *Portugal na Grande Guerra*, 1934–38, pp. 131–235.

19 MONIZ, Egas, *Um ano de política*, Lisboa, Portugal-Brasil Limitada, Sociedade Editora, 1919, pp. 295–296; FRAGA, Luís M. Alves, "Portugal na Grande Guerra. O recrutamento, a mobilização e o *roulement* no Corpo Expedicionário Português – factos e números", in *CPHM, VII Colóquio: o recrutamento militar em Portugal – Actas*, Lisboa, 1996, pp. 107–118.

20 MARTINS, Ferreira, *Portugal na Grande Guerra*, 1934–38, pp. 193–235.

21 MARTINS, Ferreira, *Portugal na Grande Guerra*, 1934–38, pp. 134–193; DUARTE, António Paulo David Silva, "Esboço para uma leitura estratégica sobre a campanha de Moçambique (1914–1918)", in *Revista Militar*, n.º 8/9 (ano 50.º), Aug./Sept., 1998, pp. 667–701.

22 ARRIFES, Marco Fortunato, *A Primeira Grande Guerra na África Portuguesa, Angola e Moçambique (1914–1918),* Lisboa, Edições Cosmos/Instituto Defesa Nacional, 2004; PIRES, Nuno Lemos, "Sobre Portugal e a Guerra na Frente Africana da Grande Guerra, 1914–1918", in *Nação e Defesa*, nº145, 2016, pp. 44–55.

23 TEIXEIRA, Nuno Severiano, *O poder e a guerra, 1914–1918. Objectivos nacionais e estratégias políticas na entrada de Portugal na Grande Guerra*, 1996, pp. 358–372.

24 FRAGA, Luís M. Alves, "Portugal na Grande Guerra. O recrutamento, a mobilização e o *roulement* no Corpo Expedicionário Português – factos e números", 1996, pp. 128–130; MARQUES, Isabel Pestana, *Os portugueses nas trincheiras. Um quotidiano de guerra*, Lisboa, CPHM, 2002, pp. 27–33.

25 MARQUES, Isabel Pestana, *Os portugueses nas trincheiras. Um quotidiano de guerra*, 2002.

26 GONÇALVES, José António Sequeira, *Sidónio Pais e a participação portuguesa na guerra (1914–1918)*, Tese de Mestrado em História Contemporânea de Portugal, Lisboa, UL, 1989; MENESES, Filipe Ribeiro de, *União Sagrada e sidonismo – Portugal na guerra (1916–18)*, 2000, pp. 259–268; FRAGA, Luís M. Alves, "Portugal na Grande Guerra. O recrutamento, a mobilização e o *roulement* no Corpo Expedicionário Português – factos e números", 1996, pp. 130–133. Mais recentemente, FRAGA, Luís M. Alves, *Do Intervencionismo ao Sidonismo. Os dois segmentos da política de guerra na 1ª República: 1916–1918*, Coimbra, Imprensa da Universidade de Coimbra, 2010.

27 MARQUES, Isabel Pestana, *Os portugueses nas trincheiras. Um quotidiano de guerra*, 2002, p. 179–287; FRAGA, Luís M. Alves, "Portugal na Grande Guerra: As deserções e os furtos no Corpo Expedicionário Português", in *Revista Militar*, n.º 5 (ano 37.º), Maio, 1985, pp. 371–395; FRAGA, Luís M. Alves, *Do Intervencionismo ao Sidonismo. Os dois segmentos da política de guerra na 1ª República: 1916–1918*, 2010, pp. 587–600.

28 FRAGA, Luís M. Alves, *La Lys: a última batalha do Exército português*, sep. *Actas do IV Colóquio: A História Militar de Portugal no Séc. XIX*, s.l., Ramos, Afonso & Moita Lda, 1993; HENRIQUES, Mendo Castro; LEITÃO, António Rosas, *La Lys – 1918 – Os soldados desconhecidos*, Lisboa, Prefácio, 2001.

29 TEIXEIRA, Nuno Severiano, "Faim et nostalgie: les prisonniers portugais de la Grande Guerre", in *Guerres Mondiales et Conflits Contemporains – Revue d'Histoire, Le Portugal dans les Conflits du XX Siècle*, n.º 178, Avr., 1995, pp. 25–41.

30 MOURA, Carneiro de, *Portugal e o Tratado de Paz*, Lisboa, Imprensa Nacional, 1918.

31 FERREIRA, José Medeiros, *Portugal na Conferência de Paz. Paris, 1919*, 1992; CRUZ, Duarte Ivo, *Estratégia Portuguesa na Conferência de Paz, 1918–1919*. Lisboa, Fundação Luso-Americana para o Desenvolvimento, 2009.

32 *Palavras claras. Razões da intervenção militar de Portugal na guerra europeia*, 1917.

4 The *Estado Novo* and the Spanish Civil War

1 ROSAS, Fernando, *O salazarismo e a aliança luso-britânica (estudos sobre a política externa do Estado Novo nos anos 30 e 40)*, Lisboa, Fragmentos, 1988.

2 TORRE GÓMEZ, Hipólito, "La conspiración iberista de Manuel Azaña", in Fernando ROSAS (coord.), *Portugal e a Guerra Civil de Espanha*, Lisboa, Edicções Colibri/IHC-UNL, 1998, pp. 209–220.

3 FARINHA, Luís, "1931: A ditadura face ao perigo espanhol", in *História*, ano XVIII (nova série), n.º 15, Dez., 1995, pp. 4–13.

4 OLIVEIRA, César, *Portugal e a II República de Espanha (1931–1936)*, Lisboa, Perspectivas & Realidades, 1985; TORRE GÓMEZ, Hipólito de

la, *A relação peninsular na antecâmara da Guerra Civil de Espanha (1931–36)*, Lisboa, Cosmos, 1998; OLIVEIRA, Pedro Aires, "Portugal e a Guerra Civil de Espanha – a retaguarda diplomática de Franco", in *História*, ano XXI (nova série), n.° 12, Mar., 1999, pp. 40–44.

5 OLIVEIRA, César, *Portugal e a II República de Espanha (1931–1936)*, 1985; TORRE GÓMEZ, Hipólito de la, *A relação peninsular na antecâmara da Guerra Civil de Espanha (1931–36), 1988*; OLIVEIRA, Pedro Aires, "Portugal e a Guerra Civil de Espanha – a retaguarda diplomática de Franco", 1999, pp. 40–44.

6 TELO, António José, "A neutralidade portuguesa na Segunda Guerra Mundial", in Nuno Severiano TEIXEIRA (coord.), *Portugal e a guerra: história das intervenções militares portuguesas nos grandes conflitos mundiais (sécs. XIX e XX)*, Lisboa, Colibri/IHC-UNL, 1998, pp. 112–113.

7 DELGADO, Iva, *Portugal e a Guerra Civil de Espanha*, Mem Martins, Europa-América, 1980, pp. 40–50; OLIVEIRA, César, *Salazar e a Guerra Civil de Espanha*, Lisboa, Editora O Jornal, 1987, pp. 303–308; OLIVEIRA, Pedro Aires, "Portugal e a Guerra Civil de Espanha – a retaguarda diplomática de Franco", 1999, pp. 40–51; GASPAR, Carlos, "Relação com a Espanha", in António BARRETO; Maria Filomena MÓNICA (coord.), *Dicionário de História de Portugal*, vol. VII, supl. A/E, Porto, Figueirinhas, 1999, pp. 650–651; ALEXANDRE, Valentim, *O roubo das Almas. Salazar, a Igreja e os totalitarismos (1930–1939)*, Lisboa, Publicações D. Quixote, 2006, pp. 195–357.

8 DELGADO, Iva, *Portugal e a Guerra Civil de Espanha*, 1980, pp. 50–70; OLIVEIRA, César, *Salazar e a Guerra Civil de Espanha*, 1987, pp. 308–327; ROSAS, Fernando, *O salazarismo e a aliança luso-britânica (estudos sobre a política externa do Estado Novo nos anos 30 e 40)*, 1988; OLIVEIRA, Pedro Aires, "Portugal e a Guerra Civil de Espanha – a retaguarda diplomática de Franco", 1999, pp. 44–51.

9 DELGADO, Iva, *Portugal e a Guerra Civil de Espanha*, 1980, pp. 81–148 and 174–188; OLIVEIRA, César, *Salazar e a Guerra Civil de Espanha*, 1987, pp. 242–281.

10 OLIVEIRA, César, *Salazar e a Guerra Civil de Espanha*, 1987.

11 ABREU, Maria Filomena, "A rádio portuguesa e a Guerra Civil de Espanha", in Fernando Rosas (coord.), *Portugal e a Guerra Civil de Espanha*, 1998, pp. 123–132.

12 OLIVEIRA, César, *Salazar e a Guerra Civil de Espanha*, 1987, p. 258.

13 *Ibidem*, p. 244.

14 *Ibidem*, pp. 270–272.

15 DELGADO, Iva, *Portugal e a Guerra Civil de Espanha*, 1980, pp. 188–202; OLIVEIRA, César, *Salazar e a Guerra Civil de Espanha*, 1987, pp. 343–366; ROSAS, Fernando, *O salazarismo e a aliança luso-britânica (estudos sobre a política externa do Estado Novo nos anos 30 e 40)*, 1988, pp. 107–120; TISCAR, Maria José, *O Pacto Ibérico a Nato e a Guerra Colonial*, Lisboa, IN-CM/Instituto Defesa Nacional, 2014, pp.

24–39; ALEXANDRE, Valentim, *O roubo das Almas. Salazar, a Igreja e os totalitarismos (1930–1939)*, 2006, pp. 195–357.

16 CARRILHO, Maria, *Forças Armadas e mudança política em Portugal no séc. XX – Para uma explicação sociológica do papel dos militares*, 1985, pp. 306–319; VIEIRA, Belchior, "Exército", in António BARRETO; Maria Filomena MÓNICA (coords.), *Dicionário de História de Portugal*, vol. VII, supl. A/E, Porto, Figueirinhas, 1999, pp. 690–692.

17 CARRILHO, Maria, *Forças Armadas e mudança política em Portugal no séc. XX – Para uma explicação sociológica do papel dos militares*, 1985, pp. 319–327; AFONSO, Aniceto; MARTELO, David, "Organização do Exército", in António BARRETO and Maria Filomena MÓNICA (coord.), *Dicionário de História de Portugal*, vol. VII, supl. A/E, Porto, Figueirinhas, 1999, pp. 695–699; DUARTE, António Paulo, *A política de Defesa Nacional, 1919–1958*, Lisboa, Imprensa de Ciências Sociais, 2010, pp. 69–120.

18 CARRILHO, Maria, *Forças Armadas e mudança política em Portugal no séc. XX – Para uma explicação sociológica do papel dos militares*, 1985, pp. 328–331; VIEIRA, Belchior, "Exército", 1999. pp. 694–695.

19 CARRILHO, Maria, *Forças Armadas e mudança política em Portugal no séc. XX – Para uma explicação sociológica do papel dos militares*, 1985, pp. 409–423; FERREIRA, José Medeiros, "As relações entre as Forças Armadas e o regime (1993–1960)", in Fernando ROSAS (coord.), *Portugal e o Estado Novo (1930–1960)*, vol. XII, in Joel SERRÃO; A. H. de Oliveira MARQUES (dir.), *Nova História de Portugal*, Lisboa, Presença, 1992; pp. 143–198; FARIA, Telmo, *Debaixo de fogo! Salazar e as Forças Armadas (1935–1941)*, Lisboa, Edições Colibri/Instituto da Defesa Nacional, 2000, pp. 165–194.

20 TELO, António José, *História da marinha portuguesa. Homens, doutrinas e organização, 1824–1974*, 1999, pp. 315–346; DUARTE, António Paulo, *A política de Defesa Nacional, 1919–1958*, 2010, pp. 39–50.

21 FARIA, Telmo, *Debaixo de fogo! Salazar e as Forças Armadas (1935–1941)*, 2000.

22 TELO, António José, "A neutralidade portuguesa na Segunda Guerra Mundial", 1998, p. 111.

23 CARRILHO, Maria, *Forças Armadas e mudança política em Portugal no séc. XX – Para uma explicação sociológica do papel dos militares*, 1985, pp. 309–310; FERREIRA, José Medeiros, "As relações entre as Forças Armadas e o regime (1993–1960)", 1992, pp. 190–192; AFONSO, Aniceto; MARTELO, David, "Organização do Exército", 1999, p. 697.

24 CARRILHO, Maria, *Forças Armadas e mudança política em Portugal no séc. XX – Para uma explicação sociológica do papel dos militares*, 1985, pp. 310–311; FERREIRA, José Medeiros, "As relações entre as Forças Armadas e o regime (1993–1960)", 1992, pp. 192–194; AFONSO, Aniceto; MARTELO, David, "Organização do Exército", 1999, pp. 695–699.

25 CARRILHO, Maria, *Forças Armadas e mudança política em Portugal no séc. XX – Para uma explicação sociológica do papel dos militares*, 1985, pp. 318–319; FERREIRA, José Medeiros, "As relações entre as Forças Armadas e o regime (1993–1960)", 1992, pp. 195–196; AFONSO, Aniceto; MARTELO, David, "Organização do Exército", 1999, p. 696; RODRIGUES, Luís Nuno, *O Corpo de Estado Maior*, Lisboa, IESM, 2015.

26 CARRILHO, Maria, *Forças Armadas e mudança política em Portugal no séc. XX – Para uma explicação sociológica do papel dos militares*, 1985, pp. 313–317; FERREIRA, José Medeiros, "As relações entre as Forças Armadas e o regime (1993–1960)", 1992, pp. 186–199.

27 TELO, António José, "As relações peninsulares num período de guerras globais (1935–1945)", in Fernando ROSAS (coord.), *Portugal e a Guerra Civil de Espanha*, 1998, p. 136.

28 VIEIRA, Belchior, "Exército", 1999, pp. 694–695.

29 TELO, António José, "A aeronáutica à espera da oportunidade", in BARATA, Manuel Themudo; Nuno Severiano TEIXEIRA, *Nova História Militar de Portugal*, vol. IV, Lisboa, Círculo de Leitores, 2004, p. 419.

30 FARIA, Telmo, *Debaixo de fogo! Salazar e as Forças Armadas (1935–1941)*, 2000.

5 The *Estado Novo* and the Second War

1 TELO, António José, "A neutralidade portuguesa na Segunda Guerra Mundial", 1998, pp. 122–125.

2 DELGADO, Iva, *Portugal e a Guerra Civil de Espanha*, 1980, pp. 188–202; OLIVEIRA, César, *Salazar e a Guerra Civil de Espanha*, Lisboa, Editora O Jornal, 1987, pp. 343–366; ROSAS, Fernando, *O salazarismo e a aliança luso-britânica (estudos sobre a política externa do Estado Novo nos anos 30 e 40)*, 1988, pp. 107–120; TISCAR, Maria José, *O Pacto Ibérico a Nato e a Guerra Colonial*, 2014, pp. 24–39.

3 TELO, António José, "A neutralidade portuguesa na Segunda Guerra Mundial", 1998, pp. 109–125.

4 DELGADO, Iva, *Portugal e a Guerra Civil de Espanha*, 1980, pp. 188–202; OLIVEIRA, César, *Salazar e a Guerra Civil de Espanha*, 1987, pp. 343–366; TELO, António José, *Portugal na Segunda Guerra*, Lisboa, Perspectivas & Realidades, 1987, pp. 209–216; ROSAS, Fernando, *O salazarismo e a aliança luso-britânica (estudos sobre a política externa do Estado Novo nos anos 30 e 40)*, 1988, pp. 107–120; TISCAR, Maria José, *O Pacto Ibérico a Nato e a Guerra Colonial*, 2014, pp. 30–39.

5 TELO, António José, *Portugal na Segunda Guerra*, 1987.

6 TELO, António José, *Portugal na Segunda Guerra (1941–1945)*, vol. 1, Lisboa, 1991, pp. 11–100.

7 FERREIRA, José Medeiros, *O comportamento político dos militares – forças armadas e regimes políticos em Portugal no século XX*, Lisboa, Estampa, 1992, p. 207; JANEIRO, Helena Pinto, *Salazar e Pétain –*

relações luso-francesas durante a Segunda Guerra Mundial (1940–44), Lisboa, Cosmos, 1998.

8 TELO, António José, *Portugal na Segunda Guerra (1941–1945)*, 1991, vol. 1, pp. 100–181; vol. 2, pp. 163–222.

9 NOGUEIRA, Franco, *História de Portugal*, II supl. in Damião PERES (dir.), *História de Portugal*, Porto, Civilização, 1981, pp. 177–224; Nogueira, NOGUEIRA, Franco, *Salazar – as grandes crises (1936–1945)*, vol. III, Porto, Civilização, 3.ª ed, 1978.

10 AAVV, *Portugal na Segunda Guerra Mundial*, Lisboa, Publicações D. Quixote, 1989; ROSAS, Fernando, *Portugal entre a paz e a guerra: Estudo do impacte da II Guerra Mundial na economia e na sociedade portuguesas (1939–1945)*, Lisboa, Estampa, 1990, pp. 131–136; TELO, António José, *Portugal na Segunda Guerra*, 1987; TELO, António José, *Portugal na Segunda Guerra (1941–1945)*, 1991; TELO, António José, *A neutralidade portuguesa e o ouro nazi*, Lisboa, Quetzal, 2000, pp. 19–80; Andrade, Luís M. Vieira de (1993) – *Neutralidade colaborante. O caso de Portugal na Segunda Guerra Mundial*, Ponta Delgada, pp. 179–211; STONE, Glyn, *The oldest ally: Britain and the Portuguese connection, 1936–1941*, London, The Royal Historical Society Woodbridge/The Boydell Press, 1994.

11 CARRILHO, Maria, *Forças Armadas e mudança política em Portugal no séc. XX– Para uma explicação sociológica do papel dos militares*, 1985, p. 320; AFONSO, Aniceto; MARTELO, David, "Organização do Exército", 1999, p. 698; RODRIGUES, Luís Nuno, *O Corpo de Estado Maior*, 2015.

12 TELO, António José, "A neutralidade portuguesa na Segunda Guerra Mundial", 1998, p. 113.

13 CARRILHO, Maria, "Política de defesa e de rearmamento", in AAVV, *Portugal na Segunda Guerra Mundial: Contributos para uma reavaliação*, Lisboa, D. Quixote, 1989, pp. 11–36; VIEIRA, Belchior, "Exército", 1999, pp. 694–695; FARIA, Telmo, *Debaixo de fogo! Salazar e as Forças Armadas (1935–1941)*, 2000, pp. 211–241.

14 TELO, António José, *História da marinha portuguesa. Homens, doutrinas e organização, 1824–1974*, 1999, pp. 369–442; DUARTE, António Paulo, *A política de Defesa Nacional, 1919–1958*, 2010, pp. 183–198.

15 FERREIRA, José Medeiros, *O comportamento político dos militares – forças armadas e regimes políticos em Portugal no século XX*, 1992, p. 203.

16 CARRILHO, Maria, *Forças Armadas e mudança política em Portugal no séc. XX – Para uma explicação sociológica do papel dos militares*, 1985, pp. 423–425; FERREIRA, José Medeiros, *O comportamento político dos militares – forças armadas e regimes políticos em Portugal no século XX*, 1992, pp. 203–222; FARIA, Telmo, *Debaixo de fogo! Salazar e as Forças Armadas (1935–1941)*, 2000.

17 TELO, António José, *Portugal na Segunda Guerra (1941–1945)*, 1991

vol. 1, pp. 79–181; FARIA, Telmo, *Debaixo de fogo! Salazar e as Forças Armadas (1935–1941)*, 2000, pp. 240–253; DUARTE, António Paulo, *A política de Defesa Nacional, 1919–1958*, 2010, pp. 157–174 and 199–212.

18 TELO, António José, *Portugal na Segunda Guerra (1941–1945)*, 1991, vol. 1, pp. 79–181; FERREIRA, José Medeiros, *O comportamento político dos militares – forças armadas e regimes políticos em Portugal no século XX*, 1992, p. 218.

19 TELO, António José, *História da marinha portuguesa. Homens, doutrinas e organização, 1824–1974*, 1999, pp. 408–441.

20 MENEZES, Manuel Amorim de Sousa, *A defesa dos Açores durante a II Guerra Mundial*, Lisboa, Estado-Maior do Exército, 1988, pp. 27–34.

21 GIANLUCA, André, *Il problema delle Azzore e la neutralitá durante la seconda guerra mondiale*, Milão, 1971; VINTRAS, Roland Eugene, *História secreta da base dos Açores*, Lisboa, Ulisseia, 1975; MENEZES, Manuel Amorim de Sousa, *A defesa dos Açores durante a II Guerra Mundial*, 1988; FERREIRA, José Medeiros, "Os Açores nas duas guerras mundiais", in *Política Internacional*, n.º 1, Jan., 1990, pp. 5–17; SANTOS, Isabel Calmeiro dos, "Os Açores e a Segunda Guerra Mundial", in *História*, ano XV, n.º 159, Dez., 1992, pp. 11–45; TELO, António José, *Os Açores e o controlo do Atlântico (1898/1948)*, 1993, pp. 289–412; TEIXEIRA, Nuno Severiano, "Le Portugal, l'Atlantique et la défense des États Unis d'Amérique pendant les deux guerres mondiales", in *XVIII Congresso Internazionale di Storia Militare*, Roma, 1993, pp. 411–414.

22 BRETES, Maria da Graça, *Timor entre invasores 1941–1945*, Lisboa, Livros Horizonte, 1989; BESSA, Carlos Gomes, "Le Portugal neutre: La Seconde Guerre mondiale 1939–1945. Le Timor portugais dans la Deuxième Guerre mondiale. Martyres et luttes en territoire neutre", in *Guerres Mondiales et Conflits Contemporains – Revue d'Histoire, Le Portugal dans les Conflits du XX Siècle*, n.º 178, 1995, pp. 43–62; SANTA, José Duarte, *Australianos e japoneses em Timor na II Guerra Mundial (1941–1945)*, Lisboa, Ed. Notícias, 1997; CARVALHO, Manuel de Abreu Ferreira, *Relatório dos Acontecimentos de Timor (1942–45)*, Lisboa, Edições Cosmos/Instituto da Defesa Nacional, 2003; MOTTA, Carlos Teixeira da, *O caso de Timor na II Guerra Mundial. Documentos Britânicos*, Lisboa, Instituo Diplomático, s.d.

6 The Cold War and NATO

1 SILVA, Isabel Alarcão, "A admissão de Portugal na ONU", in *História*, ano XV, n.º 165, Lisboa, Jun., 1993, pp. 13–37; CASTAÑO, David, "Portugal e a ONU, a primeira aproximação", in *Relações Internacionais*, n.º 47, Outubro, 2015, pp. 9–38.

2 ROLLO, Fernanda, *Portugal e o Plano Marshall*, Lisboa, Estampa, 1994.

3 VINTRAS, Roland Eugene, *História secreta da base dos Açores*, Lisboa, Ulisseia, 1975; MARQUINA, Barrio A., *España en la politica de*

seguridad del Ocidente, Madrid, Ed. del Ejército, 1986, pp. 943–960; MAGALHÃES, J. Calvet de, "Portugal e os Estados Unidos: relações no domínio da defesa", in *EREI*, n.° 3, 1987, pp. 25–29; TELO, António José, *Os Açores e o controlo do Atlântico (1898/1948)*, Porto, Asa, 1993.

4 VINTRAS, Roland Eugene, *História secreta da base dos Açores*, 1975; MARQUINA, Barrio A., *España en la politica de seguridad del Ocidente*, 1986, pp. 943–960; TELO, António José, *Os Açores e o controlo do Atlântico (1898/1948)*, 1993.

5 CARVALHO, Henrique Martins de, *Portugal e o Pacto do Atlântico*, Lisboa, Ed. Gráfica Portuguesa, 1953.

6 NOGUEIRA, Franco, *Salazar. O ataque (1945–1958)*, vol. IV, Coimbra, Atlântida, 1980; NOGUEIRA, Franco, *História de Portugal*, II supl. in Damião PERES (dir.), *História de Portugal*, Porto, Civilização, 1981, pp. 226–228.

7 NOGUEIRA, Albano, "The pull of the continent – Portugal votes for a European as well as an Atlantic role", in André de STAERKE, *NATO's anxious birth*, New York, 1985, pp. 68–75

8 FERREIRA, José Medeiros, "As ditaduras ibéricas e a fundação da Aliança Atlântica", in AAVV, *O Estado Novo das origens ao fim da autarcia (1926–1959)*, vol. I, Lisboa, Fragmentos, 1987, pp. 395–401; FERREIRA, José Medeiros, *Um século de problemas. As relações luso-espanholas da União Ibérica à Comunidade Europeia*, Lisboa, Livros Horizonte, 1989; FERREIRA, José Medeiros, "A ratificação portuguesa do Tratado do Atlântico Norte – Nota introdutória", in *Política Internacional*, vol. 1, n.° 1, Jan., 1990, pp. 155–157.

9 TEIXEIRA, Nuno Severiano, "Da neutralidade ao alinhamento: Portugal na fundação do Pacto do Atlântico", in Análise Social, vol. XXviii n.° 120 (1.°), 1993, pp. 55–80.

10 See the speeches "Miséria e medo características do momento actual", of 25 November 1947, and "O Ocidente em face da Rússia", of 28 April 1948, in SALAZAR, António de Oliveira, "Miséria e medo características do momento actual – discurso de 25 de Novembro de 1947", in *Discursos e notas políticas*, vol. VI, Coimbra, Coimbra Ed., 1950, pp. 287–311 and SALAZAR, António de Oliveira, "O Ocidente em face da Rússia, de 28 de Abril de 1948", in *Discursos e notas políticas*, vol. VI, Coimbra, Coimbra Ed., 1950, pp. 325–337.

11 See the exchange of diplomatic notes between London, Washington and Lisbon, particularly the Portuguese pro-memoria of 31 December 1948, the English and North American *aide-mémoire* of 10 January of 1949, and the Portuguese memorandum of 8 March 1949, in AHDMNE – Arquivo Histórico-Diplomático do Ministério dos Negócios Estrangeiros, Lisboa, Negócios Políticos, armário 50, maço 37.

12 See the English pro-memoria of 17 March 1949, in AHDMNE – Arquivo Histórico-Diplomático do Ministério dos Negócios Estrangeiros, Lisboa, Negócios Políticos, armário 50, maço 37.

13 See the letter by Nicolás Franco, Spanish Ambassador in Lisbon, dated

the 20 March 1949, in AHDMNE – Arquivo Histórico-Diplomático do Ministério dos Negócios Estrangeiros, Lisboa, Negócios Políticos, armário 50, maço 37.

14 See telegrams 78, 79 and 80 of the Portuguese Embassy in London dated the de 19 March 1949, Dean Achson's letter dated the 22 March 1949, in AHDMNE – Arquivo Histórico-Diplomático do Ministério dos Negócios Estrangeiros, Lisboa, Negócios Políticos, armário 50, maço 37.

15 NOGUEIRA, Franco, *Salazar. O ataque (1945–1958)*, 1980, pp. 143–144.

16 FERREIRA, José Medeiros, "As ditaduras ibéricas e a fundação da Aliança Atlântica", 1987, pp. 395–401; JIMÉNEZ REDONDO, Juan Carlos, *Franco e Salazar: As relações luso-espanholas durante a Guerra Fria*, Lisboa, Assírio & Alvim, 1996, pp. 163–191; JIMÉNEZ REDONDO, Juan Carlos, "Portugal, Espanha e a formação da NATO", in *Política Internacional*, vol. 3, n.° 19, 1999, pp. 97–111.

17 Avante!, March 1949, p. 1; Avante!, April 1949, pp. 1–2; and Avante!, May 1949, p. 2.

18 A República, 19 and 25 March 1949, pp. 4 and 1, respectively.

19 RABY, Dawn L., *Fascism and resistance in Portugal*, Manchester/New York, Manchester University Press, 1988, pp. 163–204.

20 VINTRAS, Roland Eugene, *História secreta da base dos Açores*, 1975; MARQUINA, Barrio A., *España en la politica de seguridad del Ocidente*, 1986, pp. 943–960; MAGALHÃES, José Calvet de, "Portugal e a integração europeia", in *EREI*, n.° 4, 1987/1988, pp. 31–32; TELO, António José, *Portugal e a NATO. O reencontro da tradição atlântica*, Lisboa, Cosmos, 1996, pp. 113–131.

21 CARRILHO, Maria, *Forças Armadas e mudança política em Portugal no séc. XX – Para uma explicação sociológica do papel dos militares*, Lisboa, IN-CM, 1985, pp. 377–399; FERREIRA, José Medeiros, *O comportamento político dos militares – forças armadas e regimes políticos em Portugal no século XX*, Lisboa, Estampa, 1992, pp. 255–262; AFONSO, Aniceto; MARTELO, David, "Organização do Exército", in António BARRETO and Maria Filomena MÓNICA (coord.), *Dicionário de História de Portugal*, vol. VII, supl. A/E, Porto, Figueirinhas, 1999, pp. 698–699.

22 About Portugal's commitments with NATO, see, with respect to the Army and the independent division, MACEDO, Ernesto Ferreira de (introd.), *Subsídios para o estudo do esforço militar português na década de cinquenta. Os compromissos com a O.T.A.N.*, 2 vols., Lisboa, Estado-Maior do Exército, 1988; ROCHA, Alfredo Ramos, "Portugal e a defesa ocidental: problemas militares e estratégicos (1945–1978)", in *Proserpina*, n.° 8, Mérida, Abr., 1989, pp. 71–73. Sobre as Forças Armadas no seu conjunto e a política de defesa de Portugal, TELO, António José, *Portugal e a NATO. O reencontro da tradição atlântica*, 1996, pp. 225–288.

23 VIEIRA, Belchior, "Exército", in António BARRETO; Maria Filomena

MÓNICA (coord.), *Dicionário de História de Portugal*, vol. VII, supl. A/E, Porto, Figueirinhas, 1999, pp. 691–692; AFONSO, Aniceto; MARTELO, David, "Organização do Exército", 1999, pp. 698–699.

24 TELO, António José, *História da marinha portuguesa. Homens, doutrinas e organização, 1824–1974*, tomo I, Lisboa, AM, 1999, pp. 467–487.

25 TELO, António José, *Portugal e a NATO. O reencontro da tradição atlântica*, 1996, pp. 273–274.

26 About the armament, see CARRILHO, Maria, *Forças Armadas e mudança política em Portugal no séc. XX – Para uma explicação sociológica do papel dos militares*, 1985, pp. 332–335; TELO, António José, *Portugal e a NATO. O reencontro da tradição atlântica*, 1996, pp. 225–288; TELO, António José, *História da marinha portuguesa. Homens, doutrinas e organização, 1824–1974*, 1999, pp. 478–494. About the creation of infra-structures, see ROCHA, Alfredo Ramos, "Portugal e a defesa ocidental: Problemas militares e estratégicos (1945–1978)", 1989, annex b, pp. 71–81.

27 CARRILHO, Maria, *Forças Armadas e mudança política em Portugal no séc. XX – Para uma explicação sociológica do papel dos militares*, 1985, pp. 377–399; FERREIRA, José Medeiros, *O comportamento político dos militares – forças armadas e regimes políticos em Portugal no século XX*, 1992, pp. 223 250.

28 MARTINS, Fernando, *Portugal e a Organização das Nações Unidas: Uma história política externa e ultramarina portuguesa no pós-guerra (Agosto de 1941–Setembro de 1968)*, Tese de Mestrado em História dos Sécs. XIX e XX, Lisboa, UNL, 1995; SILVA, A. E. Duarte, "O litígio entre Portugal e a ONU (1960–1974)", in *Análise Social*, vol. XXX, n.º 130, 1995, pp. 5–50; MAGALHÃES, José Calvet de, *Portugal e as Nações Unidas: a questão colonial (1955–1974)*, Lisboa, Cadernos do Lumiar 6, IEEI, 1995; CASTAÑO, David, "Portugal e a ONU, a primeira aproximação", 2015, pp. 9–38.

29 ANTUNES, José Freire, *Kennedy e Salazar. O Leão e a Raposa*, Lisboa, Difusão Cultural, 1991; RODRIGUES, Luís Nuno, *Salazar-Kennedy: A crise de uma aliança*, Lisboa, Ed. Notícias, 2002; PINTO, António Costa, *O fim do Império Português: A cena internacional, a guerra colonial, e a descolonização, 1961–1975*, Lisboa, Livros Horizonte, 2001.

30 AMARAL, Diogo Freitas do, *A tentativa falhada de um acordo Portugal–EUA sobre o futuro do ultramar português (1963)*, Coimbra, Coimbra Ed., 1994; RODRIGUES, Luís Nuno, *Salazar-Kennedy: a crise de uma aliança*, 2002, pp. 292–301.

31 MAGALHÃES, José Calvet de, "Portugal e a integração europeia", 1987–1988, pp. 32–33.

32 MAHONEY, Richard D., *J.F.K. ordeal in Africa*, New York, Oxford University Press, 1983, p. 187 and ff.; ANTUNES, José Freire, *Kennedy e Salazar. O Leão e a Raposa*, 1991, pp. 175–176 and 207–227;

RODRIGUES, Luís Nuno, *Salazar-Kennedy: A crise de uma aliança*, 2002, pp. 54–59.

33 ESTADO-MAIOR DO EXÉRCITO, *Resenha histórico-militar das campanhas de África (1961–1974)*, vol. 1, Lisboa, EME, 1988/1989.

34 ANTUNES, José Freire, *Kennedy e Salazar. O Leão e a Raposa*, 1991, pp. 239–243.

35 NATO/OTAN (1981) – *Structures faits et chifres*, Bruxelas, p. 52; DELMAS, Claude, *L'OTAN*, PUF, Paris, 1981, pp. 60–65.

36 ROCHA, Alfredo Ramos, "Portugal e a defesa ocidental: Problemas militares e estratégicos (1945–1978)", 1989, pp. 72–74.

37 MAGALHÃES, José Calvet de, "Portugal e a integração europeia", 1987–1988, p. 36; ANTUNES, José Freire, *Os americanos e Portugal. Os anos de Richard Nixon (1969–1974)*, Lisboa, D. Quixote, 1986, pp. 152–154.

38 PEREIRA, Bernardo Futcher, "Continuidade na mudança. As relações entre Portugal e os Estados Unidos", in *Política Internacional*, n.º 2, vol. 1, June 1990, pp. 5–19.

39 ANTUNES, José Freire, *Portugal na guerra do petróleo. Os Açores e as vitórias de Israel*, Carnaxide, Edeline, 2000, pp. 41–101.

40 THEMIDO, João Hall, *Dez anos em Washington (1971–1981). As verdades e os mitos nas relações luso-americanas*, Lisboa, D. Quixote, 1995, pp. 111–147, especially p. 131.

7 The End of the Empire and the Decolonisation Wars

1 FERRO, Marc, *História das colonizações: Das conquistas às independências; sécs. XIII–XX*, Lisboa, Estampa, 1996.

2 MARQUINA, Barrio A., *España en la politica de seguridad del Ocidente*, 1986.

3 DORES, Manuel António Peixeiro das, *A queda da Índia*, Tese de Mestrado em História Social e Contemporânea, Lisboa, ISCTE, 2003; STOCKER, Maria Manuel, *Xeque-Mate a Goa*, Lisboa, Temas e Debates, 2005; BÈGUE, Sandrine, *La fin de Goa et de l'Estado da Índia: Décolonization et Guerre Froide dans le Sous-Continent Indien (1945–1962)*, 2 volumes, Lisboa, MNE/Instituto Diplomático, 2007.

4 PINTO, António Costa, *O fim do Império Português: A cena internacional, a guerra colonial, e a descolonização, 1961–1975*, 2001, p. 16.

5 JIMÉNEZ REDONDO, Juan Carlos, *Franco e Salazar: As relações luso-espanholas durante a Guerra Fria*, 1996, pp. 179–227.

6 SCHNEIDMAN, Witney W., *Engaging Africa: Washington and the Fall of Portugal's Colonial Empire*, New York, University Press of America, 2004.

7 VALENÇA, Fernando, *As Forças Armadas e as crises nacionais. A Abrilada de 1961*, Lisboa, Europa-América, s.d.

8 MAHONEY, Richard D., *J.F.K. Ordeal in Africa*, 1983, p. 187 and ff.; ANTUNES, José Freire, *Kennedy e Salazar. O Leão e a Raposa*, 1991, pp. 175–176 and 207–277; and, especially, RODRIGUES, Luís Nuno,

Salazar–Kennedy: A crise de uma aliança, 2002; SCHNEIDMAN, Witney W., *Engaging Africa: Washington and the Fall of Portugal's Colonial Empire*, 2004.

9 PINTO, António Costa, *O fim do Império Português: A cena internacional, a guerra colonial, e a descolonização, 1961–1975*, 2001, pp. 18–19.

10 MAGALHÃES, José Calvet de, "Portugal e a integração europeia", 1987/1988, pp. 32–33.

11 AMARAL, Diogo Freitas do, *A tentativa falhada de um acordo Portugal–EUA sobre o futuro do ultramar português (1963)*, 1994; PINTO, António Costa, *O fim do Império Português: a cena internacional, a guerra colonial, e a descolonização, 1961–1975*, 2001, p. 25; Rodrigues, RODRIGUES, Luís Nuno, *Salazar–Kennedy: A crise de uma aliança*, 2002, p. 292 and ff.

12 MARCOS, Daniel da Silva Costa, *Salazar e de Gaulle: A França e a Questão Colonial Portuguesa*, Lisboa, MNE/Instituto Diplomático, 2007; FONSECA, Ana Mónica, *As Forças Armadas: o Apoio da República Federal da Alemanha ao Estado Novo*, Lisboa, MNE/Instituto Diplomático, 2007.

13 BELLER, Dennis Charles, *The portuguese territories issues in the United Nations: An analysis of debates*, Ph. D. dissertation, Los Angeles, University of California, 1970; SILVA, A. E. Duarte, "O litígio entre Portugal e a ONU (1960–1974)", 1995, pp. 5–50; MAGALHÃES, José Calvet de, *Portugal e as Nações Unidas: A questão colonial (1955–1974)*, 1995; MARTINS, Fernando, "A política externa do Estado Novo, o ultramar e a ONU: Uma doutrina histórico-jurídica (1955–68)", in *Penélope*, n.° 18, 1998, pp. 189 206.

14 PINTO, António Costa, *O fim do Império Português: A cena internacional, a guerra colonial, e a descolonização, 1961–1975*, 2001, p. 22.

15 COKER, Christopher, *NATO, the Warsaw Pact and Africa*, London, MacMillan, 1985.

16 MARCOS, Daniel da Silva Costa, *Salazar e de Gaulle: A França e a Questão Colonial Portuguesa*, 2007.

17 FONSECA, Ana Mónica, *As Forças Armadas: O Apoio da República Federal da Alemanha ao Estado Novo*, 2007.

18 BARROSO, Luís Machado, *Salazar e Ian Smith. O Apoio de Portugal à Rodésia (1964–1968)*, Lisboa, MNE/Instituto Diplomático, 2009.

19 PINTO, António Costa; TEIXEIRA, Nuno Severiano (ed.), *A Europa do Sul e a Construção da União europeia, 1945–2000*, Lisboa, Imprensa de Ciências Sociais, 2005, pp. 30–33; CASTILHO, José Manuel Tavares, *A ideia de Europa no marcelismo (1968–1974)*, Lisboa, Assembleia da República/Afrontamento, 2000; MAGALHÃES, J. Calvet, *Portugal na Europa: o caminho certo*, Lisboa, Bertrand Editora, 1997; LEITÃO, Nicolau Andresen, *Estado Novo, Democracia e Europa (1947–1986)*, Lisboa, Imprensa de Ciências Sociais, 2007, pp. 148–167.

20 PINTO, António Costa, *O fim do Império Português: A cena interna-*

cional, a guerra colonial, e a descolonização, 1961–1975, 2001, p. 31; MACQUEEN, Norrie, *A descolonização da África portuguesa: A revolução metropolitana e a dissolução do Império*, Mem Martins, Inquérito, 1998, pp. 78–79.

21 TEIXEIRA, Nuno Severiano, "Portugal e a NATO: 1949–1999", in *Nação e Defesa*, 2.ª série, n.º 89, 1999, p. 29.

22 ANTUNES, José Freire, *Portugal na guerra do petróleo. Os Açores e as vitórias de Israel*, 2000, pp. 41–101.

23 THEMIDO, João Hall, *Dez anos em Washington (1971–1981). As verdades e os mitos nas relações luso-americanas*, 1995, pp. 111–147, particularly p. 131.

24 SÁ, Tiago Moreira de, *História das Relações Portugal–EUA (1776–2015)*, Lisboa, D. Quixote, 2016.

25 ESTADO-MAIOR DO EXÉRCITO, *Resenha histórico-militar das campanhas de África (1961–1974)*, 1988/1989, pp. 240–241 and 259–261.

26 *Ibidem*, p. 258.

27 *Ibidem*, pp. 260–261; CANN, John p. , *Contra-insurreição em África. O modo português de fazer a guerra 1961–1974*, Lisboa, Atena, 1998, pp. 121–129.

28 AFONSO, Aniceto; MARTELO, David, "Organização do Exército", 1999, p. 699; NOGUEIRA, António Inácio, *Os Capitães do Fim . . . Do Quarto Império*, Lisboa, Âncora Editora, 2016, pp. 102–124.

29 CANN, John p. , *Contra-insurreição em África. O modo português de fazer a guerra 1961–1974*, 1998, p. 149.

30 VIEIRA, Belchior, "Exército", 1999, pp. 693–694; AFONSO, Aniceto; MARTELO, David, "Organização do Exército", 1999, p. 699.

31 MENSURADO, Joaquim M., *Os páras na guerra*, Lisboa, Prefácio, 2002; CARMO, António; MACHADO, Miguel, *Tropas pára-quedistas*, Lisboa, Prefácio, 2003, pp. 25–37.

32 CANN, John p. , *Contra-insurreição em África. O modo português de fazer a guerra 1961–1974*, 1998, pp. 133–144.

33 FREIRE, João, *Homens em fundo azul-marinho. Ensaio de observação sociológica sobre uma corporação nos meados do séc. XX: A Armada portuguesa*, Oeiras, Celta, 2003, pp. 161–168.

34 FREIRE, João, *Homens em fundo azul-marinho. Ensaio de observação sociológica sobre uma corporação nos meados do séc. XX: A Armada portuguesa*, 2003, pp. 179–184; SACHETTI, António Emílio, "A acção da Armada nas Campanhas de África", in AAVV, *Estudos sobre as campanhas de África"*, Lisboa, Atena//IAEM, 2000, pp. 205–223.

35 TELO, António José, *História da marinha portuguesa. Homens, doutrinas e organização, 1824–1974*, 1999, pp. 515–566; FREIRE, João, *Homens em fundo azul-marinho. Ensaio de observação sociológica sobre uma corporação nos meados do séc. XX: A Armada portuguesa*, 2003, pp. 155–161.

36 CORBAL, Aurélio B. Aleixo, "O vector aéreo nas campanhas de África.

Análise conceptual e estrutural", in Adriano MOREIRA; Manuel Jacinto NUNES; Francisco BALSEMÃO [et al.], *Estudos sobre as campanhas em África*, Lisboa, Atena/IAEM, 2000, pp. 189–202.

37 TAVARES, João Carlos Moreira, *A indústria militar portuguesa no tempo de guerra, 1961–1974*, Tese de Mestrado em História dos Séculos XIX e XX Lisboa, UNL, 2002.

38 CARRILHO, Maria, *Forças Armadas e mudança política em Portugal no séc. XX – Para uma explicação sociológica do papel dos militares*, 1985, pp. 331–335; TELO, António José, *Portugal e a NATO. O reencontro da tradição atlântica*, 1996, pp. 225–288; VIEIRA, Belchior, "Exército", 1999, p. 695; and, especially, TAVARES, João Carlos Moreira, *A indústria militar portuguesa no tempo de guerra, 1961–1974*, 2002, pp. 56–134.

39 TELO, António José, *História da marinha portuguesa. Homens, doutrinas e organização, 1824–1974*, 1999, pp. 556–566.

40 TELO, António José, *Portugal e a NATO. O reencontro da tradição atlântica*, 1996, pp. 224–228.

41 MACQUEEN, Norrie, *A descolonização da África portuguesa: A revolução metropolitana e a dissolução do Império*, 1998, p. 73.

42 CANN, John p. , *Contra-insurreição em África. O modo português de fazer a guerra 1961–1974*, 1998, pp. 63–89; GARCIA, Francisco Miguel Proença, "Os doutrinadores da subversão e a sua aplicabilidade nas campanhas de África (1961–1974)", in *Revista Militar*, n.° 12 (ano 50.°), Dez., 1998, pp. 1073–1121; TAVARES, João, "Forças Armadas e contra-insurreição: a guerra subversiva – nova doutrina", in *História*, ano XXI (nova série), n.° 20, Dez., 1999, pp. 48–55; FERNANDES, António Horta, "A estratégia entra na idade adulta: do pós-guerra à contra-subversão em África", in BARATA, Manuel Themudo; Nuno Severiano TEIXEIRA, *Nova História Militar de Portugal*, vol. IV, Lisboa, Círculo de Leitores, 2004, pp. 570–574.

43 FERRO, Marc, *História das colonizações: Das conquistas às independências; sécs. XIII–XX*, 1996.

44 MARTINS, Helder, *Casa dos Estudantes do Império. Subsídios para a História do seu Período mais Decisivo*, 1953–1961, Lisboa, Editorial Caminho, 2017; CASTELO, Cláudia; JERÓNIMO, Miguel Bandeira, *Casa dos Estudantes do Império. Dinâmicas Coloniais, Conexões Transnacionais*, Lisboa, Edições 70, 2017

45 MATEUS, Dalila Cabrita, *A luta pela independência: A formação das elites fundadoras da FRELIMO, MPLA e PAIGC*, Lisboa, Inquérito, 1999; PINTO, António Costa, *O fim do Império Português: A cena internacional, a guerra colonial, e a descolonização, 1961–1975*, 2001, pp. 36–44; MACQUEEN, Norrie, *A descolonização da África portuguesa: A revolução metropolitana e a dissolução do Império*, 1998, pp. 37–42; AFONSO, Aniceto, "A guerra colonial", in João Medina (dir.), *História de Portugal dos tempos pré-históricos aos nossos dias*, vol. XIII, Alfragide, Clube Internacional do Livro, 1986, pp. 337–339.

46 MATEUS, Dalila Cabrita, *A luta pela independência: A formação das elites fundadoras da FRELIMO, MPLA e PAIGC*, 1999, p. 86.
47 See, with respect to all of them, MATEUS, Dalila Cabrita, *A luta pela independência: A formação das elites fundadoras da FRELIMO, MPLA e PAIGC*, 1999.
48 NUNES, António Pires, *Angola, 1966–1974, vitória militar no Leste*, Lisboa, Prefácio, 2002.
49 AFONSO, Aniceto, "A guerra colonial", 1986, pp. 342–343.
50 MACQUEEN, Norrie, *A descolonização da África portuguesa: A revolução metropolitana e a dissolução do Império*, 1998, p. 58.
51 About the overall balance of the military situation in Angola, ESTADO-MAIOR DO EXÉRCITO, *Resenha histórico-militar das campanhas de África (1961–1974)*, 1988/1989, pp. 105–113; BARATA, Manuel Themudo, "Le Portugal et les conflits de la décolonisation: 1961–1974", in *Guerres Mondiales et Conflits Contemporains – Revue d'Histoire, Le Portugal dans les Conflits du XX Siècle*, n.° 178, 1995, pp. 79–83; and, above all,, BERNARDO, Henrique Gomes, *Estratégia de um conflito. Angola 1961–1974*, Tese de Mestrado em Estratégia, Lisboa, ISCTE, 2002.
52 For the phases of guerilla warfare, see NOGUEIRA, João de Sá, *O Exército na guerra subversiva*, 5 vols., Lisboa, Estado-Maior do Exército, 1963, pp. 9–14.
53 MACQUEEN, Norrie, *A descolonização da África portuguesa: A revolução metropolitana e a dissolução do Império*, 1998, p. 59, note 60.
54 See the chapter on Spínola in PIRES, Nuno Lemos, *Wellington, Spínola e Petraeus. O Comando Holístico da Guerra*, Lisboa, Edição Nexo Literário, 2014.
55 GARCIA, Francisco Miguel Proença, *Guiné: 1963–1974: Os movimentos independentistas, o Islão e o poder português*, Lisboa, CPHM/UPIH, 2000, pp. 155–205.
56 MACQUEEN, Norrie, *A descolonização da África portuguesa: A revolução metropolitana e a dissolução do Império*, 1998, p. 58.
57 VILLAS-BOAS, José Manuel, *Caderno de memórias*, Lisboa, Temas & Debates, 2003, pp. 100–111.
58 GOMES, Carlos de Matos, *Moçambique 1970: Operação Nó Górdio*, Lisboa, Prefácio, 2002.
59 MIDDLEMAS, Keith, *Cabora Bassa: Engineering and politics in Southern Africa*, London, Weidenfeld and Nicolson, 1975.
60 JARDIM, Jorge, *Moçambique, terra queimada*, Lisboa, Intervenção, 1976; ANTUNES, José Freire, *José Jorge Jardim, o agente secreto*, Venda Nova, Bertrand, 1996.
61 MACQUEEN, Norrie, *A descolonização da África portuguesa: A revolução metropolitana e a dissolução do Império*, 1998, p. 58.
62 PÉLISSIER, René, *Le naufrage des caravelles. Études sur la fin de l'empire portugais (1961–1975)*, Orgeval, Editions Pélissier, 1979, p. 64.
63 PÉLISSIER, René, *Le naufrage des caravelles. Études sur la fin de l'empire portugais (1961–1975)*, 1979, p. 64.

64 About the overall balance of the military situation in Mozambique, ESTADO-MAIOR DO EXÉRCITO, *Resenha histórico-militar das campanhas de África (1961–1974)*, 1988/1989, pp. 113–116; BARATA, Manuel Themudo, "Le Portugal et les conflits de la décolonisation: 1961–1974", 1995, pp. 83–89; and especially GARCIA, Francisco Miguel Proença, *Guiné: 1963–1974: Os movimentos independentistas, o Islão e o poder português*, 2000.

65 CANN, John p. , *Contra-insurreição em África. O modo português de fazer a guerra 1961–1974*, 1998, p. 248.

8 From War Campaigns to Peacekeeping Operations

1 SÁNCHEZ CERVELLÓ, Josep, *Decolonizacion y surgimento del Terceiro Mundo*, Barcelona, Hipótesi, 1997; MACQUEEN, Norrie, *A descolonização da África portuguesa: A revolução metropolitana e a dissolução do Império*, Mem Martins, Inquérito, 1998; PINTO, António Costa, *O fim do Império Português: A cena internacional, a guerra colonial, e a descolonização, 1961–1975*, Lisboa, Livros Horizonte, 2001; COSTA, João Paulo; RODRIGUES, José Damião; OLIVEIRA, Pedro Aires (coord.), "Uma Descolonização fora de Horas (1961–1975)", in *História da Expansão e do Império Português*, Lisboa, Esfera dos Livros, 2014.

2 FERREIRA, José Medeiros, "Portugal", in Charles ZORGBIBE (coord.), *Dicionário de política internacional*. Lisboa, D. Quixote, 1990, pp. 438–444; TELO, António José, "As relações internacionais da transição", in J. M. Brandão de BRITO (coord.), *Do marcelismo ao fim do Império*, Lisboa, Ed. Notícias, 1999, pp. 225–266; TEIXEIRA, Nuno Severiano; HERMENEGILDO, Reinaldo Saraiva, "As Presidências portuguesas da União Europeia", in *Relações Internacionais*, Dezembro, 2015, pp. 57–78.

3 See PCP's positions regarding NATO in *Dossier NATO (1979)*, Ed. "Avante!"; and CARVALHO, Rogério de (coord.), *Portugal e a NATO*, Lisboa, Ed. "Avante!", 1990.

4 VASCONCELOS, Álvaro de, "Los tres rostros del atlantismo portugués", in *Prosepina*, n.° 8, Mérida, Abr., 1989, pp. 129–141; SÁ, Tiago Moreira de, *Os Estados Unidos da América e a Democracia Portuguesa (1974–1976)*, Lisboa, MNE/Instituto Diplomático, 2009; SÁ, Tiago Moreira de, *História das Relações Portugal–EUA (1776–2015)*, Lisboa, D. Quixote, 2016, pp. 507–520.

5 THEMIDO, João Hall, *Dez anos em Washington (1971–1981). As verdades e os mitos nas relações luso-americanas*, Lisboa, D. Quixote, 1995, p. 167 and ff.

6 MAGALHÃES, J. Calvet de, "Portugal e os Estados Unidos: Relações no domínio da defesa", in *EREI*, n.° 3, 1987, pp. 43–51; STENHOUSE, Mark; GEORGE, Bruce, "Defense policy and strategic importance: The western perspective", in Kenneth MAXWELL (ed.) – *Portuguese defense*

an foreign policy since democratization, Camões Center, Special Report n.°3, 1991, pp. 38–41.

7 TEIXEIRA, Nuno Severiano, "Portugal e a NATO: 1949–1999", in *Nação e Defesa*, 2.ª série, n.° 89, 1999, pp. 32–33; TELO, António José, "Portugal e a NATO: Um pequeno poder numa grande aliança", in José Medeiros FERREIRA (coord.), *Política externa e política de defesa do Portugal democrático – II Curso Livre de História Contemporânea*, Lisboa, Colibri/Fundação Mário Soares/IHC-UNL, 2001, pp. 119–122.

8 MACDONALD, Scott, *European destiny, Atlantic transformation – Portuguese foreign policy under the Second Republic, 1974–1992*, Transaction Pub, 1993, pp. 65–87; TEIXEIRA, Nuno Severiano; PINTO, António Costa (ed.), *Europeanization of Portuguese Democracy*, New York, Columbia University Press, 2012, pp. 7–26.

9 BARRETO, António, "Portugal: democracy towards Europe", in Jeffrey J. ANDERSON, *Regional integration and democracy: Expanding on the European experience*, Lanham, Rowman y Littlefield, 1999, pp. 95–122; LOPES, José da Silva (ed.), *Portugal and EC membership evaluated*, London/New York, Pinter Publishers/St. Martins's Press, 1993; RIBEIRO, Maria Manuela Tavares; MELO, António Moreira Barbosa de; PORTO, Manuel Carlos Lopes (org.), *Portugal e a construção europeia*, Coimbra, Almedina, 2003; LÃ, João Rosa; CUNHA, Alice (org.), *Memórias da Adesão. À mesa das negociações*, Lisboa, Bookbuilders, 2016.

10 LAMEGO, José, "A emergência da CPLP e as suas consequências", in José Medeiros FERREIRA (coord.), *Política externa e política de defesa do Portugal democrático – II Curso Livre de História Contemporânea*, Lisboa, Colibri/Fundação Mário Soares/IHC-UNL, 2001, pp. 89–93; AAVV, *A Comunidade dos Países de Língua Portuguesa*, Lisboa, Edições Atena/Instituto de Altos Estudos Militares, 1999; PINTO, José Filipe, *Do Império Colonial à Comunidade dos Países de Língua Portuguesa: Continuidades e Descontinuidades*, Lisboa, MNE/Instituto Diplomático, 2005; MARCHUETA, Maria Regina, *A CPLP e seu enquadramento*, Lisboa, MNE/Instituto Diplomático, 2003.

11 TEIXEIRA, Nuno Severiano, "A europeização da política externa portuguesa", in *Actas do Colóquio Diplomatas e Diplomacia*, Lisboa, Livros Horizonte, 2003.

12 TEIXEIRA, Nuno Severiano, "A União Europeia como actor internacional: A política externa da União Europeia", in *Nação e Defesa*, 2.ª série, n.° 85, 1998, pp. 128–135.

13 TEIXEIRA, Nuno Severiano, "Rivalidade, subsidiariedade, complementaridade: que modelo para a identidade europeia de segurança e defesa", in *Europa Novas Fronteiras*, n.° 3, Jun., 1998, pp. 68–73.

14 TEIXEIRA, Nuno Severiano; HERMENEGILDO, Reinaldo Saraiva, "As Presidências portuguesas da União Europeia", 2015, pp. 61–78.

15 VASCONCELOS, Álvaro de, "Portugal pressing for an open Europe", in Christopher HILL (ed.), *The actors in Europe's foreign policy*,

London/New York, Routledge, 1996, pp. 268–288; TEIXEIRA, Nuno Severiano, "A europeização da política externa portuguesa", 2003.

16 BARROSO, José M. Durão, "A identidade europeia de segurança e defesa – a evolução da posição portuguesa", in *Nação e Defesa*, n.º 84, 1998, pp. 119–128; VITORINO, António, "A política de defesa nacional", in *Nação e Defesa*, ano XXIi, n.º 83, Jul./Set., 1998, pp. 40 41; VITORINO, António, *Opções de política de defesa nacional*, Lisboa, MDN, 1998, pp. 29–36 and pp. 49–54; GAMA, Jaime, "A política externa", in *Nação e Defesa*, ano XXII, n.º 83, Jul./Set., 1997, pp. 49–51.

17 About the Portuguese official position at the intergovernmental conference, see MINISTÉRIO DOS NEGÓCIOS ESTRANGEIROS, *Portugal e a Conferência Intergovernamental para a Revisão do Tratado da União Europeia*, Lisboa, MNE/IDi, 1996, pp. 39–41.

18 SERRA, Narciso Será i, *El progresso de la politica de defesa europeia*, Barcelona, Documentos CIDOB, n.º 1, 2001; NAVARRO GONZÁLEZ, Alberto, "La política europea de seguridad y defesa después de Niza", in Gustavo Palomares LERMA (ed.), Valencia, Inst. Univ. Gen. Gutiérrez Mellado, 2002, pp. 105–114.

19 GREVI, Giovanni; Helly, Damien; Keohane, Daniel, *ESDP: The first 10 years (1999–2009)*, Paris, EU ISS, 2009.

20 TEIXEIRA, Nuno Severiano, "O Tratado de Lisboa e a segurança europeia", in *Relações Internacionais*, Junho, 2008, pp. 83–90.

21 BRITO, Nuno, "Para a Acta: O Acordo de Cooperação e Defesa entre Portugal e os Estados Unidos da América", in *Política Internacional*, n.º 11, 1995, pp. 135–357; SÁ, Tiago Moreira de, *História das Relações Portugal-EUA (1776–2015)*, Lisboa, D. Quixote, 2016, p. 569.

22 TEIXEIRA, Nuno Severiano, "A NATO nas mãos de Kafka ou a metamorfose de uma aliança", in *Política Internacional*, vol. 3, n.º 19, 1999, pp. 63–69.

23 GAMA, Jaime, "A política externa", 1997, pp. 51–52.

24 TEIXEIRA, Nuno Severiano, "A Política de Defesa entre Atlantismo e Europeísmo: Portugal, a NATO e a União Europeia", in Teresa Ferreira Rodrigues; Rafael García Pérez, *Portugal e Espanha: Crise e Convergência na União Europeia*, Lisboa, Tribuna, 2011, pp. 257–266.

25 TEIXEIRA, Nuno Severiano, "Rivalidade, subsidiariedade, complementaridade: Que modelo para a identidade europeia de segurança e defesa", 1998, pp. 68–73.

26 VITORINO, António, *Opções de política de defesa nacional*, 1998, pp. 52, 69 and 70; GAMA, Jaime, "A política externa", 1997, pp. 51–52.

27 TEIXEIRA, Nuno Severiano, "A NATO nas mãos de Kafka ou a metamorfose de uma aliança", 1999, pp. 63–69.

28 TEIXEIRA, Nuno Severiano, "A Política de Defesa entre Atlantismo e Europeísmo: Portugal, a NATO e a União Europeia", 2011, pp. 265–266.

29 MINISTÉRIO DA DEFESA NACIONAL, *Participação das Forças*

Armadas portuguesas em operações de paz e na cooperação técnico-militar, súmula n.° 62, Lisboa, MDN, 1999, pp. 10–20.

30 PINTO, Alípio Tomé, "A cooperação militar na formação das Forças Armadas angolanas", in José Medeiros FERREIRA (coord.), *Política externa e política de defesa do Portugal democrático – II Curso Livre de História Contemporânea*, Lisboa, Colibri/Fundação Mário Soares/IHC-UNL, 2001, pp. 219–231; RIBEIRO, Gonçalves, *O Reencontro. Da Ponte Aérea a Cooperação*, Lisboa, Inquérito, 2006.

31 MINISTÉRIO DA DEFESA NACIONAL, *Participação das Forças Armadas portuguesas em operações de paz e na cooperação técnico-militar*, 1999, pp. 10–20.

32 DUARTE, Susana Alexandre, "A Cooperação Técnico-Militar Portuguesa", in *Revista Militar*, n°2511, 2011; MINISTÉRIO DA DEFESA NACIONAL, *Anuário Estatístico da Defesa Nacional*, 2010. Available at: http://www.portugal.gov.pt/pt/ministerios/mdn/quero-saber-mais/sobre-o-ministerio/20120227–anuarios-estatisticos/20120227–anuarios-estatisticos.aspx; PIRES, Rui, "O Exército como um vector de apoio à política externa do Estado português. A Cooperação Técnico-Militar", in Adriano MOREIRA (dir.), *Portugal e as Operações de Paz. Uma Visão Multidimensional*, Lisboa, Prefácio/Fundação Mário Soares, 2010, pp. 643–662.

33 SANTOS, J. Loureiro dos, *Forças Armadas em Portugal*, Lisboa, Fundação Francisco Manuel dos Santos, 2012, p. 71.

34 ASSEMBLEIA DA REPÚBLICA, *Programas relativos à aquisição de Equipamentos Militares*, Divisão de Informação Legislativa e Parlamentar, Lisboa, 2014, pp. 11–12.

35 BRUNEAU, Thomas, "A Relação Civil-Militar em Portugal: O Longo Processo para o Controle Civil e a Eficácia Militar", in *Revista Dados*, vol. 59, n°2, 2016, pp. 427–447.

36 SANTOS, J. Loureiro dos, *Forças Armadas em Portugal*, 2012, p. 103.

37 MINISTÉRIO DA DEFESA NACIONAL, *Anuário Estatístico da Defesa Nacional*, Lisboa, 2010.

38 For the period between 1991 and 2000, see DUQUE, Jorge, "Portugal em operações de paz – 2002", in *O mundo em português*, Abr., 2003, pp. 30; MINISTÉRIO DA DEFESA NACIONAL, *Participação das Forças Armadas portuguesas em operações de paz e na cooperação técnico-militar*, 1999, pp. 22–35. For the period between 2000 and 2010, the estimate was made based on the *Anuários Estatísticos da Defesa Nacional* 2003–2010 and the average expenditure stands at 59,702,194.50 euros. MINISTÉRIO DA DEFESA NACIONAL, *Anuário Estatístico da Defesa Nacional*, available at: http://www.portugal.gov.pt/pt/ministerios/mdn/quero-saber-mais/sobre-o-ministerio/20120227-anuarios-estatisticos/20120227-anuarios-estatisticos.aspx.

39 HELENA Carreiras, *Inquérito sobre Defesa e Forças Armadas*, Lisboa, Relatório de investigação, CIES-IUL, 2010.

40 HELENA Carreiras, *Inquérito sobre Defesa e Forças Armadas*, 2010.

41 DUQUE, Jorge, "Portugal nas missões de paz", in *Janus 98*, supl. esp. *As Forças Armadas portuguesas no novo contexto internacional*, Lisboa, Público/UAL, 1998, pp. 6–47; MINISTÉRIO DA DEFESA NACIONAL, *Participação das Forças Armadas portuguesas em operações de paz e na cooperação técnico-militar*, 1999, pp. 22–35.

42 DUQUE, Jorge, "Portugal nas missões de paz", 1998, pp. 46–47; MINISTÉRIO DA DEFESA NACIONAL, *Participação das Forças Armadas portuguesas em operações de paz e na cooperação técnico-militar*, 1999, pp. 22–35.

43 BRITO, Villa de (coord.), *Bósnia 96: Missão do Exército em Apoio à Operação de Paz*, Lisboa, Secção de Informação Protocolo e Relações Públicas do Gabinete do General Chefe de Estado-Maior do Exército, 1997.

44 DUQUE, Jorge, "Portugal nas missões de paz", 1998, pp. 46–47; MINISTÉRIO DA DEFESA NACIONAL, *Participação das Forças Armadas portuguesas em operações de paz e na cooperação técnico-militar*, 1999, pp. 22–35.

45 TEIXEIRA, Nuno Severiano, "Das campanhas de guerra às operações de paz: as intervenções militares portuguesas em teatro europeu", in *EREI*, n.º 14, 2.º semestre, 1999, pp. 21–32.

46 For a global view, see MOREIRA, Adriano (dir.), *Portugal e as Operações de Paz. Uma Visão Multidimensional*, Lisboa, Prefácio/Fundação Mário Soares, 2010. For a thorough view of the international missions, see BRANCO, Carlos Martins, "As Forças Armadas, Forças de Segurança e a Política Externa do Estado", in *Anuário de Política Externa Portuguesa*, Lisboa, IPRI/MNE (in press).

47 CARREIRAS, Helena, "O que pensam os militares portugueses do "peacekeeping"?", in *EREI*, n.º 14, 2.º semestre, 1999, pp. 65–95; CARRILHO, Maria "Os conflitos nos Balcãs e a redefinição das missões internacionais", in José Medeiros FERREIRA (coord.), *Política externa e política de defesa do Portugal democrático – II Curso Livre de História Contemporânea*, Lisboa, Colibri/Fundação Mário Soares/IHC-UNL, 2001, pp. 182–185; MATOS, Luís Salgado; Bacalhau, Mário, *As Forças Armadas em tempo de mudança. Uma sondagem à opinião pública nas vésperas do século xxi*, Lisboa, Cosmos/IDN, 2001, pp. 120–170.

General and Specialised Bibliography

GENERAL BIBLIOGRAPHY

BARATA, Manuel Themudo; TEIXEIRA, Nuno Severiano, *Nova História Militar de Portugal*, Lisboa, Círculo de Leitores, 2004.

BARRETO, António; MÓNICA, Maria Filomena (coord.), *Dicionário de História de Portugal*, Porto, Figueirinhas, 1999.

BETHENCOURT, Francisco; CHAUDHURI, Kirti, *História da Expansão Portuguesa*, Lisboa, Círculo de Leitores, 1998.

COSTA, João Paulo; RODRIGUES, José Damião; OLIVEIRA, Pedro Aires (coord.), *História da Expansão e do Império Português*, Lisboa, Esfera dos Livros, 2014.

MACEDO, Jorge, *História Diplomática Portuguesa – constantes e linhas de força*, Lisboa, Instituto Defesa Nacional, 1987.

MATTOSO, José, *História de Portugal*, Lisboa, Círculo de Leitores, 1993.

MEDINA, João (dir.), *História de Portugal dos tempos pré-históricos aos nossos dias*, Alfragide, Clube Internacional do Livro, 1986.

PERES, Damião (dir.), *História de Portugal*, Barcelos, Portucalense, 1933–1981.

RENOUVIN, Pierre, *Histoire des Relations Internationales*, Paris, Hachete, 1994.

SERRÃO, Joel; MARQUES, A. H. Oliveira (dir.), *Nova História de Portugal*, Lisboa, Presença, 1992.

SOUSA, Bernardo Vasconcelos e; RAMOS, Rui; MONTEIRO, Nuno Gonçalo, *História de Portugal*, Lisboa, Esfera dos Livros, 2009.

TELO, António José, *História da marinha portuguesa. Homens, doutrinas e organização, 1824–1974*, Lisboa, AM, 1999.

ZORGBIBE, Charles (coord.), *Dicionário de política internacional*. Lisboa, D. Quixote, 1990.

SPECIALISED BIBLIOGRAPHY

AAVV, *Portugal na Segunda Guerra Mundial*, Lisboa, Publicações D. Quixote, 1989.

ABREU, Maria Filomena, "A rádio portuguesa e a Guerra Civil de Espanha", in Fernando ROSAS (coord.), *Portugal e a Guerra Civil de Espanha*, Lisboa, Colibri/IHC-UNL, 1998.

AAVV, *A Comunidade dos Países de Língua Portuguesa*, Lisboa, Edições Atena/Instituto de Altos Estudos Militares, 1999.

AFONSO, Aniceto, "A guerra colonial", in João Medina (dir.), *História de Portugal dos tempos pré-históricos aos nossos dias*, vol. XIII, Alfragide, Clube Internacional do Livro, 1986.

AFONSO, Aniceto; MARTELO, David, "Organização do Exército", in António BARRETO e Maria Filomena MÓNICA (coord.), *Dicionário de História de Portugal*, vol. VII, supl. A/E, Porto, Figueirinhas, 1999.

AFONSO, Aniceto, *Grande Guerra, Angola, Moçambique e Flandres, 1914–1918*, Lisboa, Quidnovi, 2008.

ALEXANDRE, Valentim, *Origens do Colonialismo Português Moderno*, Lisboa, Sá da Costa, 1979.

ALEXANDRE, Valentim, "Nação e Império", in Francisco BETHENCOURT; Kirti CHAUDHURI, *História da Expansão Portuguesa*, vol. IV, Lisboa, Círculo de Leitores, 1998.

ALEXANDRE, Valentim, "Situações Coloniais: II – O Ponto de Viragem: as campanhas de ocupação (1890–1930)", in Francisco BETHENCOURT; Kirti CHAUDHURI, *História da Expansão Portuguesa*, vol. IV, Lisboa, Círculo de Leitores, 1998.

ALEXANDRE, Valentim; Dias, Jill, "O Império Africano 1825–1890", tomo X, in Joel SERRÃO; A. H. Oliveira MARQUES (dir.), *Nova História de Portugal*, Lisboa, Editorial Estampa, 1998.

ALEXANDRE, Valentim (coord.), "O Império africano (séculos XIX–XX) – As linhas gerais", in *Império Africano – séculos XIX–XX*, Lisboa, Edições Colibri, 2000.

ALEXANDRE, Valentim, *O roubo das Almas. Salazar, a Igreja e os totalitarismos (1930–1939)*, Lisboa, Publicações D. Quixote, 2006.

ALMADA, José de, *Tratados Aplicáveis ao Ultramar*, Lisboa, vol. IV, 1943.

AMARAL, Diogo Freitas do, *A tentativa falhada de um acordo Portugal–EUA sobre o futuro do ultramar português (1963)*, Coimbra, Coimbra Ed., 1994.

ANDRADE, Luís M. Vieira de, *Neutralidade colaborante. O caso de Portugal na Segunda Guerra Mundial*, Ponta Delgada, s.e, 1993.

ANTUNES, José Freire, *Os americanos e Portugal. Os anos de Richard Nixon (1969–1974)*, Lisboa, D. Quixote, 1986.

ANTUNES, José Freire, *Kennedy e Salazar. O Leão e a Raposa*, Lisboa, Difusão Cultural, 1991.

ANTUNES, José Freire, *José Jorge Jardim, o agente secreto*, Venda Nova, Bertrand, 1996.

ANTUNES, José Freire, *Portugal na guerra do petróleo. Os Açores e as vitórias de Israel*, Carnaxide, Edeline, 2000.

ASSEMBLEIA DA REPÚBLICA, *Programas relativos à aquisição de Equipamentos Militares*, Divisão de Informação Legislativa e Parlamentar, Lisboa, 2014.

ARRIFES, Marco Fortunato, *A Primeira Guerra Mundial na África portuguesa, Angola e Moçambique*, Dissertação de Mestrado em História, Lisboa, FCSH-UNL, 2002.

ARRIFES, Marco Fortunato, *A Primeira Grande Guerra na África Portuguesa, Angola e Moçambique (1914–1918)*, Lisboa, Edições Cosmos/Instituto Defesa Nacional, 2004.

AXELSON, Eric, *Portugal and the Scrumble for Africa 1875–1891*, Joanesburgo, Witwatersrand University, 1967.

BARATA, Manuel Themudo, "Le Portugal et les conflits de la décolonisation: 1961–1974", in *Guerres Mondiales et Conflits Contemporains – Revue d'Histoire, Le Portugal dans les Conflits du XX Siècle*, n.° 178, 1995, pp. 79–83.

BARRETO, António, "Portugal: democracy towards Europe", in Jeffrey J. ANDERSON, *Regional integration and democracy: expanding on the European experience*, Lanham, Rowman y Littlefield, 1999.

BARROSO, José M. Durão, "A identidade europeia de segurança e defesa – a evolução da posição portuguesa", in *Nação e Defesa*, n.° 84, 1998, pp. 119–128.

BARROSO, Luís Machado, *Salazar e Ian Smith. O Apoio de Portugal à Rodésia (1964–1968)*, Lisboa, MNE/Instituto Diplomático, 2009.

BEBIANO, Rui, "Organização e Papel do Exército", in José MATTOSO, *História de Portugal*, Lisboa, Círculo de Leitores, 1993.

BÈGUE, Sandrine, *La fin de Goa et de l'Estado da Índia: Décolonization et Guerre Froide dans le Sous-Continent Indien (1945–1962)*, 2 volumes, Lisboa, MNE/Instituto Diplomático, 2007.

BELLER, Dennis Charles, *The portuguese territories issues in the United Nations: an analysis of debates*, Ph. D. dissertation, Los Angeles, University of California, 1970.

BERNARDO, Henrique Gomes, *Estratégia de um conflito. Angola 1961–1974*, Tese de Mestrado em Estratégia, Lisboa, ISCTE, 2002.

BESSA, Carlos Gomes, "Le Portugal neutre: la Seconde Guerre mondiale 1939–1945. Le Timor portugais dans la Deuxième Guerre mondiale. Martyres et luttes en territoire neutre", in *Guerres Mondiales et Conflits Contemporains – Revue d'Histoire, Le Portugal dans les Conflits du XX Siècle*, n.° 178, 1995, pp. 43–62.

BESSA, Carlos, "Angola, do feudalismo luso-africano à ocupação", in Manuel Themudo BARATA; Nuno Severiano TEIXEIRA, *Nova História Militar de Portugal*, vol. III, Lisboa, Círculo de Leitores, 2004.

BESSA, Carlos, "Guiné, das feitorias isoladas ao "enclave" unificado", in Manuel Themudo BARATA; Nuno Severiano TEIXEIRA, *Nova História Militar de Portugal*, vol. III, Lisboa, Círculo de Leitores, 2004.

BESSA, Carlos, "Moçambique, Ocupação Efectiva Peculiar", in Manuel Themudo BARATA; Nuno Severiano TEIXEIRA, *Nova História Militar de Portugal*, vol. III, Lisboa, Círculo de Leitores, 2004.

BESSA, Carlos, "Timor. Do domínio liurai à pacificação portuguesa", in Manuel Themudo BARATA; Nuno Severiano TEIXEIRA, *Nova História Militar de Portugal*, vol. III, Lisboa, Círculo de Leitores, 2004.

BONIFÁCIO, Fátima, *Seis Estudos sobre o Liberalismo português*, Lisboa, Editorial Estampa, 1991.

BONIFÁCIO, Maria de Fátima, *História da Guerra Civil da Patuleia, 1846–1847*, Lisboa, Editorial Estampa, 1993.

BONIFÁCIO, Fátima, "Relações externas e Revolução Liberal, Séc. XVIII–XIX", in João Marques de ALMEIDA; Rui RAMOS, *Revoluções, Política Externa e Política de Defesa em Portugal, Séc. XIX–XX*, Lisboa, Edições Cosmos/Instituto Defesa Nacional, 2008.

BONIFÁCIO, Maria de Fátima, *Um Homem Singular. Biografia Política de Rodrigo da Fonseca Magalhães, 1787–1858*, Lisboa, D. Quixote, 2013.

BOUVIER, Jean, *Initiation au Vocabulaire et aux Mécanismes Économiques Contemporains,* Paris, SEDES, 1977.

BRANCO, Carlos Martins, "As Forças Armadas, Forças de Segurança e a Política Externa do Estado", in *Anuário de Política Externa Portuguesa*, Lisboa, IPRI/MNE (no prelo).

BRANDÃO, Raul, *El-Rei Junot*, Lisboa, IN-CM, 1982.

BRETES, Maria da Graça, *Timor entre invasores 1941–1945*, Lisboa, Livros Horizonte, 1989.

BRITO, Nuno, "Para a Acta: o Acordo de Cooperação e Defesa entre Portugal e os Estados Unidos da América", in *Política Internacional*, n.° 11, 1995, pp. 135–357.

BRITO, Villa de (coord.), *Bósnia 96: Missão do Exército em Apoio à Operação de Paz*, Lisboa, Secção de Informação Protocolo e Relações Públicas do Gabinete do General Chefe de Estado-Maior do Exército, 1997.

BRUNSCHWIG, Henri, *Le Partage de l'Afrique Noire,* Paris, Flammarion, 1971.

BRUNEAU, Thomas, "A Relação Civil-Militar em Portugal: O Longo Processo para o Controle Civil e a Eficácia Militar", in *Revista Dados*, vol. 59, n°2, 2016, pp. 427–447.

CABRAL, Manuel Villaverde, "A Grande Guerra e o sidonismo (esboço interpretativo)", in *Análise Social*, vol. XV, n.° 58, 1979, pp. 373–392.

CAETANO, Marcello, *Tradições, Princípios e Métodos da Colonização Portuguesa,* Lisboa, Agência do Ultramar, 1951.

CAETANO, Marcello, *Portugal e a Internacionalização dos Problemas Africanos,* Lisboa, Ática, 1971.

CANN, John p. , *Contra-insurreição em África. O modo português de fazer a guerra 1961–1974*, Lisboa, Atena, 1998.

CARMO, António; MACHADO, Miguel, *Tropas pára-quedistas*, Lisboa, Prefácio, 2003.

CARREIRAS, Helena, "O que pensam os militares portugueses do "peace-keeping"?", in *EREI*, n.° 14, 2.° semestre, 1999, pp. 65–95.

CARRILHO, Maria, *Forças Armadas e mudança política em Portugal no séc. XX – Para uma explicação sociológica do papel dos militares*, Lisboa, IN-CM, 1985.

CARRILHO, Maria, "Política de defesa e de rearmamento", in AAVV,

Portugal na Segunda Guerra Mundial: contributos para uma reavaliação, Lisboa, D. Quixote, 1989.

CARRILHO, Maria "Os conflitos nos Balcãs e a redefinição das missões internacionais", in José Medeiros FERREIRA (coord.), *Política externa e política de defesa do Portugal democrático – II Curso Livre de História Contemporânea*, Lisboa, Colibri/Fundação Mário Soares/IHC-UNL, 2001.

CARVALHO, Henrique Martins de, *Portugal e o Pacto do Atlântico*, Lisboa, Ed. Gráfica Portuguesa, 1953.

CARVALHO, Manuel de Abreu Ferreira, *Relatório dos Acontecimentos de Timor (1942–45)*, Lisboa, Edições Cosmos/Instituto da Defesa Nacional, 2003.

CARVALHO, Rogério de (coord.), *Portugal e a NATO*, Lisboa, Ed. "Avante!", 1990.

CASTAÑO, David, "Portugal e a ONU, a primeira aproximação", in *Relações Internacionais*, nº 47, Outubro, 2015, pp. 9–38.

CASTELO, Cláudia, *O Modo Português de Estar no Mundo. O Luso-Tropicalismo e a Ideologia Colonial Portuguesa (1933–1961)*, Lisboa, Edições Afrontamento, 1998.

CASTELO, Cláudia/ JERÓNIMO, Miguel Bandeira, *A casa dos Estudantes do Império. Dinâmicas coloniais, conexões transnacionais*, Lisboa, Edições 70, 2017.

CASTILHO, José Manuel Tavares, *A ideia de Europa no marcelismo (1968–1974)*, Lisboa, Assembleia da República/Afrontamento, 2000.

CHAGAS, Pinheiro, *Um Reinado Trágico*, Lisboa, vol. I, 1908.

CIDADE, Hernâni, "Portugal na Guerra Mundial: 1914–1918", in Damião PERES (dir.), *História de Portugal*, vol. VII, Barcelos, Portucalense, 1933.

CLARENCE-SMITH, Gervase, *The Third Portuguese Empire*, Manchester, Manchester University Press, 1985.

COKER, Christopher, *NATO, the Warsaw Pact and Africa*, Londres, MacMillan, 1985.

CORBAL, Aurélio B. Aleixo, "O vector aéreo nas campanhas de África. Análise conceptual e estrutural", in Adriano MOREIRA; Manuel Jacinto NUNES; Francisco BALSEMÃO [et al.], *Estudos sobre as campanhas em África*, Lisboa, Atena/IAEM, 2000.

CORDEIRO, Luciano, *Obras de Luciano Cordeiro: questões coloniais*, vol. 1, Coimbra, 1934.

COSTA, Fernando, *Portugal e a Guerra Anglo-Bóer: política externa e opinião pública (1899–1902)*, Lisboa, Cosmos, 1998.

CRUZ, Duarte Ivo, *Estratégia Portuguesa na Conferência de Paz, 1918–1919*. Lisboa, Fundação Luso-Americana para o Desenvolvimento, 2009.

DELGADO, Iva, *Portugal e a Guerra Civil de Espanha*, Mem Martins, Europa-América, 1980.

DELMAS, Claude, *L'OTAN*, PUF, Paris, 1981.

DEROU, Jean, *Les relations Franco-Portugaises 1910–1926*, Paris, Publications de la Sorbonne, 1986.

DORES, Manuel António Peixeiro das, *A queda da Índia*, Tese de Mestrado em História Social e Contemporânea, Lisboa, ISCTE, 2003.
Dossier NATO (1979), Ed. "Avante!"
DUARTE, António Paulo David Silva, "Esboço para uma leitura estratégica sobre a campanha de Moçambique (1914–1918)", in *Revista Militar*, n.° 8/9 (ano 50.°), Ag./Set., 1998, pp. 667–701.
DUARTE, António Paulo, *A política de Defesa Nacional, 1919–1958*, Lisboa, Imprensa de Ciências Sociais, 2010.
DUARTE, Susana Alexandre, "A Cooperação Técnico-Militar Portuguesa", in *Revista Militar*, n° 2511, 2011.
DUROSELLE, Jean-Baptiste, *L'Europe de 1815 à nos Jours,* Paris, Puf, 1964.
DUQUE, Jorge, "Portugal nas missões de paz", in *Janus 98*, supl. esp. *As Forças Armadas portuguesas no novo contexto internacional*, Lisboa, Público/UAL, 1998, pp. 46–47.
DUQUE, Jorge, "Portugal em operações de paz – 2002", in *O mundo em português*, Abr., 2003, pp. 30–32.
ESTADO-MAIOR DO EXÉRCITO, *Resenha histórico-militar das campanhas de África (1961–1974)*, vol. 1, Lisboa, EME, 1988/1989.
FARIA, Telmo, *Debaixo de fogo! Salazar e as Forças Armadas (1935–1941)*, Lisboa, Edições Colibri/Instituto da Defesa Nacional, 2000.
FARINHA, Luís, "1931: a ditadura face ao perigo espanhol", in *História*, ano XVIII (nova série), n.° 15, Dez., 1995, pp. 4–13.
FERNANDES, António Horta, "A estratégia entra na idade adulta: do pós-guerra à contra-subversão em África", in Manuel Themudo BARATA; Nuno Severiano TEIXEIRA, *Nova História Militar de Portugal*, vol. IV, Lisboa, Círculo de Leitores, 2004.
FERREIRA, José Medeiros, "As ditaduras ibéricas e a fundação da Aliança Atlântica", in AAVV, *O Estado Novo das origens ao fim da autarcia (1926–1959)*, vol. I, Lisboa, Fragmentos, 1987.
FERREIRA, José Medeiros, *Um século de problemas. As relações luso-espanholas da União Ibérica à Comunidade Europeia*, Lisboa, Livros Horizonte, 1989.
FERREIRA, José Medeiros, "Os Açores nas duas guerras mundiais", in *Política Internacional*, n.° 1, Jan., 1990, pp. 5–17.
FERREIRA, José Medeiros, "A ratificação portuguesa do Tratado do Atlântico Norte – Nota introdutória", in *Política Internacional*, vol. 1, n.° 1, Jan., 1990, pp. 155–157.
FERREIRA, José Medeiros, "Portugal", in Charles ZORGBIBE (coord.), *Dicionário de política internacional.* Lisboa, D. Quixote, 1990.
FERREIRA, José Medeiros, "As relações entre as Forças Armadas e o regime (1993–1960)", vol. XII, in Joel SERRÃO e A. H. Oliveira MARQUES (dir.), *Nova História de Portugal*, Lisboa, Presença, 1992.
FERREIRA, José Medeiros, *O comportamento político dos militares – forças armadas e regimes políticos em Portugal no século XX*, Lisboa, Estampa, 1992.

FERREIRA, José Medeiros, *Portugal na Conferência de Paz. Paris, 1919*, Lisboa, Quetzal, 1992.

FERRO, Marc, *História das colonizações: das conquistas às independências; sécs. XIII–XX*, Lisboa, Estampa, 1996.

FONSECA, Ana Mónica, *As Forças Armadas: o Apoio da República Federal da Alemanha ao Estado Novo*, Lisboa, MNE/Instituto Diplomático, 2007.

FRAGA, Luís M. Alves, "Portugal na Grande Guerra: As deserções e os furtos no Corpo Expedicionário Português", in *Revista Militar*, n.º 5 (ano 37.º), Maio, 1985, pp. 371–395.

FRAGA, Luís M. Alves, *La Lys: a última batalha do Exército português*, sep. *Actas do IV Colóquio: A História Militar de Portugal no Séc. XIX*, s.l., Ramos, Afonso & Moita Lda, 1993.

FRAGA, Luís M. Alves, "Portugal na Grande Guerra. O recrutamento, a mobilização e o *roulement* no Corpo Expedicionário Português – factos e números", in *CPHM, VII Colóquio: o recrutamento militar em Portugal – Actas*, Lisboa, 1996.

FRAGA, Luís M. Alves, *O fim da ambiguidade: a estratégia nacional portuguesa de 1914 a 1916*, Lisboa, Universitária, 2001.

FRAGA, Luís M. Alves, *Do Intervencionismo ao Sidonismo. Os dois segmentos da política de guerra na 1ª República: 1916–1918*. Coimbra, Imprensa da Universidade de Coimbra, 2010.

FREIRE, João, *Homens em fundo azul-marinho. Ensaio de observação sociológica sobre uma corporação nos meados do séc. XX: a Armada portuguesa*, Oeiras, Celta, 2003.

GAMA, Jaime, "A política externa", in *Nação e Defesa*, ano XXII, n.º 83, Jul./Set., 1997, pp. 49–51.

GARCIA, Francisco Miguel Proença, "Os doutrinadores da subversão e a sua aplicabilidade nas campanhas de África (1961–1974)", in *Revista Militar*, n.º 12 (ano 50.º), Dez., 1998, pp. 1073–1121.

GARCIA, Francisco Miguel Proença, *Guiné: 1963–1974: os movimentos independentistas, o Islão e o poder português*, Lisboa, CPHM/UPIH, 2000.

GASPAR, Carlos, "Relação com a Espanha", in António BARRETO; Maria Filomena MÓNICA (coord.), *Dicionário de História de Portugal*, vol. VII, supl. A/E, Porto, Figueirinhas, 1999.

GERARD, C. L. ; LESOURD, J. A., *La nouvelle histoire economique*, vol. 1, Paris, Librairie Armand Colin, 1976.

GIANLUCA, André, *Il problema delle Azzore e la neutralitá durante la seconda guerra mondiale*, Milão, 1971.

GOMES, Carlos de Matos, *Moçambique 1970: Operação Nó Górdio*, Lisboa, Prefácio, 2002.

GONÇALVES, José António Sequeira, *Sidónio Pais e a participação portuguesa na guerra (1914–1918)*, Tese de Mestrado em História Contemporânea de Portugal, Lisboa, UL, 1989.

GREVI, Giovanni; Helly, Damien; Keohane, Daniel, *ESDP: The first 10 years (1999–2009)*, Paris, EU ISS, 2009.

GUIMARÃES, Ângela, "A ideologia colonialista em Portugal no último quartel do século XIX", in *Ler História, n.º* 1, de Janeiro/Abril, 1983, pp. 69–79.
GUIMARÃES, Ângela, *Uma Corrente do Colonialismo Português,* Lisboa, Livros Horizonte, 1984.
HAMMOND, Richard, *Portugal and Africa 1815–1910. A Study in Uneconomic Imperialism,* Stanford-California, 1966.
HELENA Carreiras, *Inquérito sobre Defesa e Forças Armadas*, Lisboa, Relatório de investigação, CIES-IUL, 2010.
HENRIQUES, Mendo Castro; LEITÃO, António Rosas, *La Lys – 1918 – Os soldados desconhecidos*, Lisboa, Prefácio, 2001.
JANEIRO, Helena Pinto, *Salazar e Pétain – relações luso-francesas durante a Segunda Guerra Mundial (1940–44)*, Lisboa, Cosmos, 1998.
JARDIM, Jorge, *Moçambique, terra queimada*, Lisboa, Intervenção, 1976.
JERÓNIMO, Miguel Bandeira, *Portugal e a Grande Guerra. Contextos e Protagonistas*, Lisboa, Edições 70, 2015.
JIMÉNEZ REDONDO, Juan Carlos, *Franco e Salazar: as relações luso-espanholas durante a Guerra Fria*, Lisboa, Assírio & Alvim, 1996.
JIMÉNEZ REDONDO, Juan Carlos, "Portugal, Espanha e a formação da NATO", in *Política Internacional*, vol. 3, n.º 19, 1999, pp. 97–111.
LÃ, João Rosa; CUNHA, Alice (org.), *Memórias da Adesão. À mesa das negociações*, Lisboa, Bookbuilders, 2016.
LAMEGO, José, "A emergência da CPLP e as suas consequências", in José Medeiros FERREIRA (coord.), *Política externa e política de defesa do Portugal democrático – II Curso Livre de História Contemporânea*, Lisboa, Colibri/Fundação Mário Soares/IHC-UNL, 2001.
LANGHORNE, Richard, "Anglo-German negotiation concerning the future of the Portuguese colonies, 1911–1914", in *The Historical Journal*, vol. XVI, n.º 2, 1973, pp. 361–387.
LEITÃO, Niclolau Andresen, *Estado Novo, Democracia e Europa (1947–1986)*, Lisboa, Imprensa de Ciências Sociais, 2007.
LOPES, Fernando Farelo, *Poder político e caciquismo na 1.ª República Portuguesa*, Lisboa, Estampa, 1994.
LOPES, José da Silva (ed.), *Portugal and EC membership evaluated*, Londres/Nova Iorque, Pinter Publishers/St. Martins's Press, 1993.
MACDONALD, Scott, *European destiny, Atlantic transformation – Portuguese foreign policy under the Second Republic, 1974–1992*, Transaction Pub, 1993.
MACQUEEN, Norrie, *A descolonização da África portuguesa: a revolução metropolitana e a dissolução do Império*, Mem Martins, Inquérito, 1998.
MACEDO, Ernesto Ferreira de (introd.), *Subsídios para o estudo do esforço militar português na década de cinquenta. Os compromissos com a O.T.A.N.*, 2 vols., Lisboa, Estado-Maior do Exército, 1988.
MAGALHÃES, J. Calvet de, "Portugal e os Estados Unidos: relações no domínio da defesa", in *EREI*, n.º 3, 1987, pp. 25–29.

MAGALHÃES, José Calvet de, "Portugal e a integração europeia", in *EREI*, n.º 4, 1987/1988, pp. 31–32.

MAGALHÃES, José Calvet de, *Portugal e as Nações Unidas: a questão colonial (1955–1974)*, Lisboa, Cadernos do Lumiar 6, IEEI, 1995.

MAGALHÃES, J. Calvet, *Portugal na Europa: o caminho certo*, Lisboa, Bertrand Editora, 1997.

MAHONEY, Richard D., *J.F.K. ordeal in Africa*, Nova Iorque, Oxford University Press, 1983.

MARCOS, Daniel da Silva Costa, *Salazar e de Gaulle: a França e a Questão Colonial Portuguesa*, Lisboa, MNE/Instituto Diplomático, 2007.

MARQUES, A. H. de Oliveira (coord.), *Portugal da Monarquia para a República*, vol. XI, in Joel SERRÃO; A. H. de Oliveira MARQUES (dir.), *Nova História de Portugal*, Lisboa, Presença, 1991.

MARQUES, Fernando Pereira, *Exército, mudança e modernização na primeira metade do século XIX*, Lisboa, Edições Cosmos/Instituto da Defesa Nacional, 1999.

MARQUES, Isabel Pestana, *Os portugueses nas trincheiras. Um quotidiano de guerra*, Lisboa, CPHM, 2002.

MARQUINA, Barrio A., *España en la politica de seguridad del Ocidente*, Madrid, Ed. del Ejército, 1986.

MARTINS, Fernando, *Portugal e a Organização das Nações Unidas: uma história política externa e ultramarina portuguesa no pós-guerra (Agosto de 1941–Setembro de 1968)*, Tese de Mestrado em História dos Sécs. XIX e XX, Lisboa, UNL, 1995.

MARTINS, Fernando, "A política externa do Estado Novo, o ultramar e a ONU: uma doutrina histórico-jurídica (1955–68)", in *Penélope*, n.º 18, 1998, pp. 189–206.

MARTINS, Ferreira, *Portugal na Grande Guerra*, Lisboa, Ática, 1934–38.

MARTINS, Hélder, *Casa dos Estudantes do Império. Subsídios para a História do seu período mais decisivo (1953 a 1961)*, Lisboa, Editorial Caminho, 2017.

MARTINS, Oliveira, *Portugal Contemporâneo*, vol. 2, Lisboa, Guimarães Editora, 1986.

MARCHUETA, Maria Regina, *A CPLP e seu enquadramento*, Lisboa, MNE/Instituto Diplomático, 2003.

MATEUS, Dalila Cabrita, *A luta pela independência: a formação das elites fundadoras da FRELIMO, MPLA e PAIGC*, Lisboa, Inquérito, 1999.

MATOS, Luís Salgado; Bacalhau, Mário, *As Forças Armadas em tempo de mudança. Uma sondagem à opinião pública nas vésperas do século XXI*, Lisboa, Cosmos/IDN, 2001.

MENESES, Filipe Ribeiro de, *União Sagrada e sidonismo – Portugal na guerra (1916–18)*, Lisboa, Cosmos, 2000.

MENEZES, Manuel Amorim de Sousa, *A defesa dos Açores durante a II Guerra Mundial*, Lisboa, Estado-Maior do Exército, 1988.

MENSURADO, Joaquim M., *Os páras na guerra*, Lisboa, Prefácio, 2002.

MIDDLEMAS, Keith, *Cabora Bassa: Engeneering and politics in Southern Africa*, Londres, Weidenfeld and Nicolson, 1975.

MIÈGE, Jean-Louis, *Expansion Européenne et décolonisation de 1870 à nous jours*, Paris, Puf, 1973.

MILZA, Pierre, *Les Relations Internationales de 1871 à 1914,* Paris, Librairie Armand Colin, 1968.

MINISTÉRIO DA DEFESA NACIONAL, *Participação das Forças Armadas portuguesas em operações de paz e na cooperação técnico-militar*, súmula n.° 62, Lisboa, MDN, 1999.

MINISTÉRIO DA DEFESA NACIONAL, *Anuário Estatístico da Defesa Nacional*, 2010. Disponível em: http://www.portugal.gov.pt/pt/ministerios/mdn/quero-saber-mais/sobre-o-ministerio/20120227-anuarios-estatisticos/20120227-anuarios-estatisticos.aspx.

MINISTÉRIO DA DEFESA NACIONAL, *Anuário Estatístico da Defesa Nacional*, Lisboa, 2010.

MINISTÉRIO DOS NEGÓCIOS ESTRANGEIROS, *Portugal e a Conferência Intergovernamental para a Revisão do Tratado da União Europeia*, Lisboa, MNE/IDi, 1996.

MONIZ, Egas, *Um ano de política*, Lisboa, Portugal-Brasil Limitada, Sociedade Editora, 1919.

MOREIRA, Adriano (dir.), *Portugal e as Operações de Paz. Uma Visão Multidimensional*, Lisboa, Prefácio/Fundação Mário Soares, 2010.

MOURA, Carneiro de, *Portugal e o Tratado de Paz*, Lisboa, Imprensa Nacional, 1918.

MOTTA, Carlos Teixeira da, *O caso de Timor na II Guerra Mundial. Documentos Britânicos*, Lisboa, Instituo Diplomático, s.d.

NATO/OTAN (1981) – *Structures faits et chifres*, Bruxelas.

NAVARRO GONZÁLEZ, Alberto, "La política europea de seguridad y defesa después de Niza", in Gustavo Palomares LERMA (ed.), *Política de seguridad de la Unión Europea: realidades y retos para el siglo XXI*, Valencia, Inst. Univ. Gen. Gutiérrez Mellado, 2002.

NIVEAU, Maurice, *Histoire des Faits Économiques Contemporains,* Paris, Presses universitaires de France, 1966.

NOGUEIRA, Albano, "*The pull of the continent – Portugal votes for an European as well as an atlantic role*", in André DE STAERCKE, *NATO' anxious birth*, Nova Iorque, 1985.

NOGUEIRA, António Inácio, *Os Capitães do Fim . . . Do Quarto Império*, Lisboa, Âncora Editora, 2016.

NOGUEIRA, Franco, *Salazar – as grandes crises (1936–1945)*, vol. III, Porto, Civilização, 3.ª ed, 1978.

NOGUEIRA, Franco, *Salazar. O ataque (1945–1958)*, vol. IV, Coimbra, Atlântida, 1980.

NOGUEIRA, Franco, *História de Portugal*, II supl. in Damião PERES (dir.), *História de Portugal*, Porto, Civilização, 1981.

NOGUEIRA, João de Sá, *O Exército na guerra subversiva*, 5 vols., Lisboa, Estado-Maior do Exército, 1963.

NUNES, António Pires, *Angola, 1966–1974, vitória militar no Leste*, Lisboa, Prefácio, 2002.

OLIVEIRA, César, *Portugal e a II República de Espanha (1931–1936)*, Lisboa, Perspectivas & Realidades, 1985.

OLIVEIRA, César, *Salazar e a Guerra Civil de Espanha*, Lisboa, Editora O Jornal, 1987.

OLIVEIRA, Pedro Aires, "Portugal e a Guerra Civil de Espanha – a retaguarda diplomática de Franco", in *História*, ano XXI (nova série), n.º 12, Mar., 1999, pp. 40–44.

COSTA, João Paulo; RODRIGUES, José Damião; OLIVEIRA, Pedro Aires (coord.), "Uma Descolonização fora de Horas (1961–1975)", in *História da Expansão e do Império Português*, Lisboa, Esfera dos Livros, 2014.

Palavras claras. Razões da intervenção militar de Portugal na guerra europeia. Relatório publicado no "Diário do Governo" n.º 9, 1.ª série de 17 de Janeiro de 1917, Lisboa, Imprensa Nacional, 1917.

PÉLISSIER, René, *Le naufrage des caravelles. Études sur la fin de l'empire portugais (1961–1975)*, Orgeval, Editions Pélissier, 1979.

PÉLISSIER, Réne, *História das Campanhas de Angola – resistência e revoltas (1945–1941)*, 2 vols., Lisboa, Editorial Estampa, 1986.

PÉLISSIER, Réne, *História de Moçambique – formação e oposição, 1854–1918*, 2 vols., Lisboa, Editorial Estampa, 1987.

PÉLISSIER, Réne, *História da Guiné – portugueses e africanos Senegâmbia (1841–1936)*, 2 vols., Lisboa, Editorial Estampa, 1989.

PÉLISSIER, Réne, *As Campanhas Coloniais de Portugal*, Lisboa, Estampa, 2004.

PEREIRA, Bernardo Futcher, "Continuidade na mudança. As relações entre Portugal e os Estados Unidos, in *Política Internacional*, n.º 2, vol. 1, Jun., 1990, pp. 5–19.

PINTO, Alípio Tomé, "A cooperação militar na formação das Forças Armadas angolanas", in José Medeiros FERREIRA (coord.), *Política externa e política de defesa do Portugal democrático – II Curso Livre de História Contemporânea*, Lisboa, Colibri/Fundação Mário Soares/IHC-UNL, 2001.

PINTO, António Costa, *O fim do Império Português: a cena internacional, a guerra colonial, e a descolonização, 1961–1975*, Lisboa, Livros Horizonte, 2001.

PINTO, António Costa; TEIXEIRA, Nuno Severiano (ed.), *A Europa do Sul e a Construção da União europeia, 1945–2000*, Lisboa, Imprensa de Ciências Sociais, 2005.

PINTO, José Filipe, *Do Império Colonial à Comunidade dos Países de Língua Portuguesa: Continuidades e Descontinuidades*, Lisboa, MNE/Instituto Diplomático, 2005.

PIRES, Nuno Lemos, *Wellington, Spínola e Petraeus. O comando holístico da guerra*, Lisboa, Nexo Literário, 2014.

PIRES, Nuno Lemos, "Sobre Portugal e a Guerra na Frente Africana da Grande Guerra, 1914–1918", in *Nação e Defesa*, nº 145, 2016, pp. 44–55.

PIRES, Rui, "O Exército como um vector de apoio à política externa do Estado português. A Cooperação Técnico-Militar", in Adriano MOREIRA (dir.), *Portugal e as Operações de Paz. Uma Visão Multidimensional*, Lisboa, Prefácio/Fundação Mário Soares, 2010.

RABY, Dawn L., *Fascism and resistance in Portugal*, Manchester/Nova Iorque, Manchester University Press, 1988.

RIBEIRO, Gonçalves, *O Reencontro. Da Ponte Aérea a Cooperação*, Lisboa, Inquérito, 2006.

RIBEIRO, Maria Manuela Tavares; MELO, António Moreira Barbosa de; PORTO, Manuel Carlos Lopes (org.), *Portugal e a construção europeia*, Coimbra, Almedina, 2003.

ROBINSON, Ronald; GALLAGHER, John; DENNY, Alice, *Africa and the Victorians: The Official Mind of Imperialism,* Londres, Macmillan, 1961.

ROCHA, Alfredo Ramos, "Portugal e a defesa ocidental: problemas militares e estratégicos (1945–1978)", in *Proserpina*, n.º 8, Mérida, Abr., 1989, pp. 71–73.

RODRIGUES, Luís Nuno, *Salazar-Kennedy: a crise de uma aliança*, Lisboa, Ed. Notícias, 2002.

RODRIGUES, Luís Nuno, *O Corpo de Estado Maior*, Lisboa, IESM, 2015.

ROLLO, Fernanda, *Portugal e o Plano Marshall*, Lisboa, Estampa, 1994.

ROSAS, Fernando, *O salazarismo e a aliança luso-britânica (estudos sobre a política externa do Estado Novo nos anos 30 e 40)*, Lisboa, Fragmentos, 1988.

ROSAS, Fernando, *Portugal entre a paz e a guerra: estudo do impacte da II Guerra Mundial na economia e na sociedade portuguesas (1939–1945)*, Lisboa, Estampa, 1990.

SÁ, Tiago Moreira de, *Os Estados Unidos da América e a Democracia Portuguesa (1974–1976)*, Lisboa, MNE/Instituto Diplomático, 2009.

SÁ, Tiago Moreira de, *História das Relações Portugal–EUA (1776–2015)*, Lisboa, D. Quixote, 2016.

SACHETTI, António Emílio, "A acção da Armada nas Campanhas de África", in AAVV, *Estudos sobre as campanhas de África*", Lisboa, Atena//IAEM, 2000.

SALAZAR, António de Oliveira, "Miséria e medo características do momento actual – discurso de 25 de Novembro de 1947", in *Discursos e notas políticas*, vol. VI, Coimbra, Coimbra Ed., 1950.

SALAZAR, António de Oliveira, "O Ocidente em face da Rússia, de 28 de Abril de 1948", in *Discursos e notas políticas*, vol. VI, Coimbra, Coimbra Ed., 1950.

SÁNCHEZ CERVELLÓ, Josep, *Decolonizacion y surgimento del Terceiro Mundo*, Barcelona, Hipótesi, 1997.

SANTA, José Duarte, *Australianos e japoneses em Timor na II Guerra Mundial (1941–1945)*, Lisboa, Ed. Notícias, 1997.

SANTOS, Isabel Calmeiro dos, "Os Açores e a Segunda Guerra Mundial", in *História*, ano XV, n.º 159, Dez., 1992, pp. 11–45.

SANTOS, J. Loureiro dos, *Forças Armadas em Portugal*, Lisboa, Fundação Francisco Manuel dos Santos, 2012.

SANTOS, M. Emília Madeira, *Viagens de Exploração Terrestre dos Portugueses em África,* Lisboa, Centro de Estudos de Cartografía Antiga, 1978.

SARDICA, José Miguel, "Os Militares e a Política entre o Ultimato e a República", in Fernando MARTINS (ed.), *Diplomacia & Guerra*, Lisboa, Edições Colibri, 2001.

SARDICA, José Miguel, "A Revolução Liberal e a política de defesa e segurança, 1820–1851", in João Marques de ALMEIDA; Rui RAMOS, *Revoluções, Política Externa e Política de Defesa em Portugal, Séc. XIX–XX*, Lisboa, Edições Cosmos/Instituto Defesa Nacional, 2008.

SCHNEIDMAN, Witney W., *Engaging Africa: Washington and the Fall of Portugal's Colonial Empire*, New York, University Press of America, 2004.

SELVAGEM, Carlos, *Portugal Militar*, Lisboa, IN-CM, 2006.

SERRA, Narciso Será i, *El progresso de la politica de defesa europeia*, Barcelona, Documentos CIDOB, n.º 1, 2001.

SILVA, A. E. Duarte, "O litígio entre Portugal e a ONU (1960–1974)", in *Análise Social*, vol. XXX, n.º 130, 1995, pp. 5–50.

SILVA, Isabel Alarcão e, "A admissão de Portugal na ONU", in *História*, ano XV, n.º 165, Lisboa, Jun., 1993, pp. 13–37.

SMITH, John Vincent, "The Anglo-German negotiation over the Portuguese colonies in Africa, 1911–1914", in *The Historical Journal*, vol. XVII, n.º 3, 1974, pp. 562–629.

SMITH, John Vincent, *As relações políticas luso-britânicas, 1910–1916*, Lisboa, Livros Horizonte, 1975.

SOARES, Mário, *Portugal amordaçado*, Lisboa, Arcádia, 1974.

STENHOUSE, Mark; GEORGE, Bruce, "Defense policy and strategic importance: the western perspective", in Kenneth MAXWELL (ed.) – *Portuguese defense an foreign policy since democratization*, Camões Center, Special Report n.º 3, 1991.

STOCKER, Maria Manuel, *Xeque-Mate a Goa*, Lisboa, Temas e Debates, 2005.

STONE, Glyn, *The oldest ally: Britain and the Portuguese connection, 1936–1941*, Londres, The Royal Historical Society, Woodbridge/The Boydell Press, 1994.

TAVARES, João, "Forças Armadas e contra-insurreição: a guerra subversiva – nova doutrina", in *História*, ano XXI (nova série), n.º 20, Dez., 1999, pp. 48–55.

TAVARES, João Carlos Moreira, *A indústria militar portuguesa no tempo de guerra, 1961–1974*, Tese de Mestrado em História dos Séculos XIX e XX Lisboa, UNL, 2002.

TEIXEIRA, Nuno Severiano, *O Ultimatum Inglês – política externa e polítca interna no Portugal de 1890*, Lisboa, Alfa, 1990.

TEIXEIRA, Nuno Severiano, "Le Portugal, l'Atlantique et la défense des États

Unis d'Amérique pendant les deux guerres mondiales", in *XVIII Congresso Internazionale di Storia Militare*, Roma, 1993.

TEIXEIRA, Nuno Severiano, "Da neutralidade ao alinhamento: Portugal na fundação do Pacto do Atlântico", in *Análise Social*, vol. XXVIII n.° 120 (1.°), 1993, pp. 55–80.

TEIXEIRA, Nuno Severiano, "Faim et nostalgie: les prisionniers portugais de la Grande Guerre", in *Guerres Mondiales et Conflits Contemporains – Revue d'Histoire, Le Portugal dans les Conflits du XX Siècle*, n.° 178, Avr., 1995, pp. 25–41.

TEIXEIRA, Nuno Severiano, *O poder e a guerra, 1914–1918. Objectivos nacionais e estratégias políticas na entrada de Portugal na Grande Guerra*, Lisboa, Estampa, 1996.

TEIXEIRA, Nuno Severiano, "Portugal no contexto internacional", in, Francisco BETHENCOURT; Kirti CHAUDHURY (dirs.), *História da Expansão Portuguesa*, vol. IV, Lisboa, Círculo de Leitores, 1998.

TEIXEIRA, Nuno Severiano, "Colónias e colonização portuguesa na cena internacional (1885–1930)", in Francisco BETHENCOURT; Kirti CHAUDHURI, *História da Expansão Portuguesa*, vol. IV, Lisboa, Círculo de Leitores, 1998.

TEIXEIRA, Nuno Severiano, "A União Europeia como actor internacional: a política externa da União Europeia", in *Nação e Defesa*, 2.ª série, n.° 85, 1998, pp. 128 135.

TEIXEIRA, Nuno Severiano, "Rivalidade, subsidiariedade, complementaridade: que modelo para a identidade europeia de segurança e defesa", in *Europa Novas Fronteiras*, n.° 3, Jun., 1998, pp. 68–73.

TEIXEIRA, Nuno Severiano, "A NATO nas mãos de Kafka ou a metamorfose de uma aliança", in *Política Internacional*, vol. 3, n.° 19, 1999, pp. 63–69.

TEIXEIRA, Nuno Severiano, "Portugal e a NATO: 1949–1999", in *Nação e Defesa*, 2.ª série, n.° 89, 1999, p. 29.

TEIXEIRA, Nuno Severiano, "Das campanhas de guerra às operações de paz: as intervenções militares portuguesas em teatro europeu", in *EREI*, n.° 14, 2.° semestre, 1999, pp. 21–32.

TEIXEIRA, Nuno Severiano, "Entre a África e a Europa: a política externa portuguesa, 1890–1986", in António Costa PINTO (coord.), *Portugal Contemporâneo*, Madrid, Sequitur, 2000.

TEIXEIRA, Nuno Severiano, "A europeização da política externa portuguesa", in *Actas do Colóquio Diplomatas e Diplomacia*, Lisboa, Livros Horizonte, 2003.

TEIXEIRA, Nuno Severiano, "O Tratado de Lisboa e a segurança europeia", in *Relações Internacionais*, Junho, 2008, pp. 83–90.

TEIXEIRA, Nuno Severiano, "A Política de Defesa entre Atlantismo e Europeísmo: Portugal, a NATO e a União Europeia", in Teresa Ferreira Rodrigues; Rafael García Pérez, *Portugal e Espanha: Crise e Convergência na União Europeia*, Lisboa, Tribuna, 2011.

TEIXEIRA, Nuno Severiano; PINTO, António Costa (ed.), *Europeanization of Portuguese Democracy*, Nova Iorque, Columbia University Press, 2012.

TEIXEIRA, Nuno Severiano; HERMENEGILDO, Reinaldo Saraiva, "As Presidências portuguesas da União Europeia", in *Relações Internacionais*, Dezembro, 2015, pp. 61–78.

TELO, António José, *O sidonismo e o movimento operário: luta de classes em Portugal, 1917–1919*, Lisboa, Ulmeiro, 1977.

TELO, António José, *Decadência e queda da I República Portuguesa*, 1.º vol., Lisboa, A Regra do Jogo, 1980.

TELO, António José, *Portugal na Segunda Guerra*, Lisboa, Perspectivas & Realidades, 1987.

TELO, António José, *Portugal na Segunda Guerra (1941–1945)*, vol. 1 e 2, Lisboa, 1991.

TELO, António José, *Lourenço Marques na política externa portuguesa 1875–1900*, Lisboa, Cosmos, 1991.

TELO, António José, *Os Açores e o controlo do Atlântico (1898/1948)*, Porto, Asa, 1993.

TELO, António José, *Portugal e a NATO. O reencontro da tradição atlântica*, Lisboa, Cosmos, 1996.

TELO, António José, "A neutralidade portuguesa na Segunda Guerra Mundial", in Nuno Severiano TEIXEIRA (coord.), *Portugal e a guerra: história das intervenções militares portuguesas nos grandes conflitos mundiais (sécs. XIX e XX)*, Lisboa, Colibri/IHC-UNL, 1998.

TELO, António José, "As relações peninsulares num período de guerras globais (1935–1945)", in Fernando ROSAS (coord.), *Portugal e a Guerra Civil de Espanha*, Lisboa, Edições Colibri/IHC-UNL, 1998.

TELO, António José, "As relações internacionais da transição", in J. M. Brandão de BRITO (coord.), *Do marcelismo ao fim do Império*, Lisboa, Ed. Notícias, 1999.

TELO, António José, *A neutralidade portuguesa e o ouro nazi*, Lisboa, Quetzal, 2000.

TELO, António José, "Portugal e a NATO: um pequeno poder numa grande aliança", in José Medeiros FERREIRA (coord.), *Política externa e política de defesa do Portugal democrático – II Curso Livre de História Contemporânea*, Lisboa, Colibri/Fundação Mário Soares/IHC-UNL, 2001.

TELO, António José, "A aeronáutica à espera da oportunidade", in Manuel Themudo BARATA; Nuno Severiano TEIXEIRA, *Nova História Militar de Portugal*, vol. IV, Lisboa, Círculo de Leitores, 2004.

TELO, António José; SOUSA, Pedro Marquês de, *O CEp. Os militares sacrificados pela má política*, Lisboa, Fronteira do Caos, 2016.

THEMIDO, João Hall, *Dez anos em Washington (1971–1981). As verdades e os mitos nas relações luso-americanas*, Lisboa, D. Quixote, 1995.

TISCAR, Maria José, *O Pacto Ibérico a Nato e a Guerra Colonial*, Lisboa, IN-CM/Instituto Defesa Nacional, 2014.

TORRE GÓMEZ, Hipólito de la, *Conspiração contra Portugal (1910–1912)*, Lisboa, 1978.

TORRE GÓMEZ, Hipólito de la, *Na encruzilhada da Grande Guerra: Portugal-Espanha, 1913–1919*, Lisboa, Estampa, 1980.

TORRE GÓMEZ, Hipólito de la, *A relação peninsular na antecâmara da Guerra Civil de Espanha (1931–36)*, Lisboa, Cosmos, 1998.

TORRE GÓMEZ, Hipólito, "La conspiración iberista de Manuel Azaña", in Fernando ROSAS (coord.), *Portugal e a Guerra Civil de Espanha*, Lisboa, Edicções Colibri/IHC-UNL, 1998.

TORRE GÓMEZ, Hipólito de la, *El Imperio del Rey. Alfonso XIII, Portugal y los Ingleses (1907–1916)*, Mérida, Editora Regional de Extremadura, Junta de Extremadura/Unión Europea, 2002.

VALENÇA, Fernando, *As Forças Armadas e as crises nacionais. A Abrilada de 1961*, Lisboa, Europa-América, s.d.

VALENTE, Vasco Pulido, "Portugal e a Guerra de 14–18 (1)", in *O Tempo e o Modo*, n.° 33, 1965, pp. 1202–1215.

VALENTE, Vasco Pulido, *O poder e o povo. A Revolução de 1910*, Lisboa, Moraes, 1982.

VALENTE, Vasco Pulido, "Revoluções: A "República Velha" (ensaio de interpretação política)", in *Análise Social*, vol. XXVII, n.° 115 (1.°), 1992, pp. 7–63.

VALENTE, Vasco Pulido, *A "República Velha" (1910–1917)*, Ensaio, Lisboa, Gradiva, 1997.

VALENTE, Vasco Pulido, *Os Militares e a Politica (1820–1856)*, Lisboa, IN-CM, 1997.

VASCONCELOS, Álvaro de, "Los tres rostros del atlantismo portugués", in *Prosepina*, n.° 8, Mérida, Abr., 1989, pp. 129–141.

VASCONCELOS, Álvaro de, "Portugal pressing for an open Europe", in Christopher HILL (ed.), *The actors in Europe's foreign policy*, Londres/Nova Iorque, Routledge, 1996.

VAZ, Nuno Mira, *Civilinização das Forças Armadas nas sociedades demoliberais*, Lisboa, Cosmos/IDN, 2002.

VENTURA, António, *Da Revolução de 1820 ao fim das Guerras Civis*, in Manuel Themudo BARATA; Nuno Severiano TEIXEIRA, *Nova História Militar de Portugal*, vol. III, Lisboa, Círculo de Leitores, 2004.

VENTURA, António, *As Guerras Liberais, 1820–1834*, Lisboa, Academia Portuguesa de História, 2008.

VIEIRA, Belchior, "Exército", in António BARRETO; Maria Filomena MÓNICA (coords.), *Dicionário de História de Portugal*, vol. VII, supl. A/E, Porto, Figueirinhas, 1999.

VILLAS-BOAS, José Manuel, *Caderno de memórias*, Lisboa, Temas & Debates, 2003.

VILHENA, Júlio de, *Antes da República*, vol. I, Lisboa, França e Arménio, 1916.

VINTRAS, Roland Eugene, *História secreta da base dos Açores*, Lisboa, Ulisseia, 1975.

VITORINO, António, "A política de defesa nacional", in *Nação e Defesa*, ano XXII, n.º 83, Jul./Set., 1998, pp. 40–41.

VITORINO, António, *Opções de política de defesa nacional*, Lisboa, MDN, 1998.

WILLEQUET, Jacques, "Anglo-German rivalry in Belgian and Portuguese Africa", in *Britain and German in Africa, imperial rivalry and colonial rule*, New Haven/Londres, 1967.

Index

Abrilada (1824), 10–11, 12
Abrilada (1961), 131, 140
Absolute Monarchy (pre-1834), 1–15
  Army reforms, 5–6, 9, 11–12
  breakdown of, 1
  colonial model, 22, 23, 28, 39
  Navy reforms, 7–8
absolutism, 4, 11, 12, 15
Abyssinia Crisis (1935), 79
Acheson, Dean, 115
Action Committee of the Cabinda National Union (CAUNC), 148
Admiralty Council, 7, 9
Afghanistan
  peace missions, 186, 190, 191–2
  Soviet invasion, 164–5
Africa
  demarcation of borders treaty (1891), 34, 37, 38, 43
  European scramble for, 26–35
  first European expeditions, 28
  First World War theatre, 41, 42, 43, 54, 60, 65–6, 67, 74
  metropole/colonies relations, 24, 25
  Portuguese economic development, 23, 24–5
  Portuguese imperialism, 22–6, 28 9, 30–4, 38–9, 54–5
  Portuguese "public work expeditions", 30–1
  Portuguese territorial occupation, 23, 24, 25, 28, 29, 33, 34, 38–9, 40–6
  Versailles Peace Conference (1919), 75
  see also Algeria; Angola; Cape Verde Islands; Guinea-Bissau; Mozambique; São Tomé e Príncipe
African elites, 146, 147, 155
African Party for the Independence of Guinea and Cape Verde (PAIGC), 139, 147, 149, 152, 153–4
African Party for the Independence (PAI), 147
African Revolutionary Front for the National Independence of the Portuguese Colonies (FRAIN), 148–9
Afro-Asian movement, 129, 132, 134, 135, 146
Agadir crisis, 51
Aguiar, Joaquim António de, 19
Air Force
  as an auxiliary weapon, 91
  autonomy of, 86, 91, 119, 121
  decolonisation wars, 142–3, 144–5, 154
  Estado Novo rearmament plans, 104, 121
  Estado Novo reforms, 143, 144–5
  First Republic, 63
  First World War, 86
  growing importance of, 90
  IBERLAND, 124, 165
  NATO military commitments, 119, 124, 165
  Second World War, 107–8
  Spanish Civil War, 83
  transfer to the Azores, 107–8
Air Transport Brigade (BAT), 165
Aire-sur-la-Lys, 68
Ajaua ethnic group, 42
Albania
  NATO membership, 174–5
  relations with Portugal, 163
Alcobaça agreement (1820), 8
Alfonso XIII, King of Spain, 49, 52
Algarve, Liberal Wars (1828–1834), 13, 14
Algeciras, Treaty of, 49
Algeria
  decolonisation war, 136, 139, 145, 160
  French colonisation, 24, 29

All African Peoples' Conference (1960), 148
Almeida, Francisco José de Lacerda e, 28
Almeida, João de, 42, 43
Ambriz, 24, 29, 32
Ambrizete, 42
Amorim, Pedro Francisco Massano de, 65
Amsterdam, Treaty of, 170
Anderson Plan, 133–4
Andrade, Mário Pinto de, 146, 147
Anglo-Boer War, 48
Anglo-Portuguese Alliance
- *Estado Novo*, 77, 81
- First World War, 56, 57, 73–4
- pre-First World War, 47, 48, 49, 52, 53
- Second World War, 93–4, 96
- significance of (pre-1851), 2, 3, 4
- Spanish Civil War, 81
- *see also* Britain

Angoche, 24, 42
Angola
- African elites, 146, 147
- Africanisation of Portuguese military, 141
- Anglo-German agreement (1898), 48, 74
- Anglo-Portuguese border treaty (1884), 31–2
- Anglo-Portuguese disputes, 29
- Anti-Colonial Movement (MAC), 148
- "bearers" decree (1856), 25
- colonial army, 37–8
- decolonisation wars, 130–1, 141, 142, 143, 150–1, 159
- ethno-nationalist movements, 147–8
- farm workers' strike (1956), 149
- First World War, 43, 54, 65–6, 74
- FRAIN membership, 149
- German territorial claims, 55
- independence movement leaders, 146
- liberation movements, 131, 147–8, 149, 150–1, 159, 180
- Meridional Africa, 28, 33–4
- Organic Charter, 37
- outbreak of war (1961), 136
- "pacification" campaigns, 39, 42–4, 45, 64
- peace missions, 186–7
- Portuguese Air regions, 143
- Portuguese early expeditions, 28
- Portuguese interior penetration zones, 22
- Portuguese naval command, 143
- Portuguese "public work expeditions", 30–1
- Portuguese special troops, 142
- Portuguese territorial commands, 142
- Portuguese territorial occupation, 23, 24, 29, 33, 38–9, 42–4
- post-colonial technical-military cooperation, 180, 181
- revolt against compulsory cotton production (1961), 149
- UN Security Council Resolution (1961), 133, 135
- WWII mobilisation, 107
- *see also* Cabinda enclave; Luanda

Angolan Communist Party (PCA), 147
Angónia, 41
Antas, Francisco Xavier da Silva Pereira, General Conde das, 20
Anti-Colonial Movement (MAC), 148
anti-colonialism, 111, 123, 127, 129, 146
Antunes, Melo, 162
Arab states, Yom Kippur War, 138
Argentina, official recognition of First Portuguese Republic (1910), 50
Armed Forces Movement (MFA) (1974), 141, 160–1, 162, 183
Army
- Absolute Monarchy reforms, 5–6, 9, 11–12
- Africanisation of troops, 142, 158, 160
- Air Transport Brigade (BAT), 165
- "casting of lots", 6, 18, 35
- civil war (1846–1847), 20–1
- colonial army structure, 37–8
- Constitutional Charter, 11–12
- Constitutional Governments, 182
- Constitutional Monarchy reforms, 18, 21, 35–8
- decolonisation wars, 140, 141, 142, 144, 158, 159–60
- defence plan (1943), 106–7
- English reorganisation of, 5–7
- *Estado Novo* rearmament plans, 90–1, 103–4, 121, 143, 144
- *Estado Novo* reforms, 89–91, 103, 120, 142

Expeditionary Corps (CEP), 67–72
First Republic reforms, 61–4
Independent Mixed Brigade (BAI), 165
Junot's decrees, 5
Liberal Revolution (1820), 8–9
Liberal Wars (1828–1834), 12–13
liberalisms in conflict, 15
liberalist period reforms, 16, 17–19, 21
Maria da Fonte Revolt (1846), 19–20
military capacity of, 36–7
military defence of the Azores, 108
military dictatorship reforms, 85–7, 90
military expenditure, 88, 91, 105, 122, 145
modernisation of weaponry, 35, 36, 38
NATO military commitments, 119, 124, 165
"pacification" campaigns, 37, 38–46, 64
pronunciamentos, 8, 12, 17, 19
recruitment system, 6, 9, 17–18, 35, 37, 62, 89
resistance against the *Vintista* government, 10
Second World War, 103–4, 105, 108
Spanish Civil War, 83
Vintista reforms, 9
see also First World War

Army Ministry, 119
Army School, 18, 64
Arriaga, Kaúlza de, 156, 157, 158
Australia
British colonisation, 24
East Timor peace mission, 191
Austria
Holy Alliance, 2–3
rearmament of Portugal, 104
Azaña, Manuel, 79
Azores
Liberal Wars (1828–1834), 13
military divisions, 18, 21, 36, 62, 86
NATO role, 113
Portuguese government transfer plan, 106
Second World War, 95, 96, 106, 107–8
strategic value of, 51, 107, 108, 113
UK military bases, 99–100, 104, 105, 106, 108
US military bases, 99, 100–1, 108, 110, 111, 113, 118, 125, 132–3, 137–8, 164–5, 173
WWII mobilisation, 107
see also Lajes Agreements

baconga ethnic group, 147
Badajoz, 81–2
BAI (Independent Mixed Brigade), 165
Bailundo war (1902), 42, 44
Baker, James, 173
balance of power
Congress of Vienna (1815), 2–3
extra-European zones, 28
Iberian Peninsula, 1–2
Portuguese colonies, 54
Balanta ethnic group, 45
Balkans
peace missions, 188–90, 191
secession conflicts, 178
Ball, George, 123, 133, 134
Baptista, Pedro João, 28
Barbosa du Bocage, José Vicente, 32
Barreto, Luiz do Rego, 10
Barué campaign, 41
Bastos, João Pereira, 62
BAT (Air Transport Brigade), 165
Belémzada, 17
Belfastada episode (1828), 13
Belgium
African imperial project, 29, 32, 43
claim to Cabinda, 75
counter-guerrilla warfare, 145
Eurocorps, 169
NATO founding member, 112
Portuguese African colonies, 54
Belmonte Fort, 44
Bembe mines, 24
Benguela, 30, 151
Beresford, William Carr, 2, 5, 6, 7, 8, 9
Berkeley, George Cranfield, 7
Berlin Blockade (1948), 112
Berlin Conference (1884), 32–3, 34, 42
Berlin Plus agreement, 172
Berlin Wall, 167
Bermuda, 93
Bevin, Ernest, 115
Bicesse Accords, 186
Bié, 28, 30, 33, 44
Bismarck, Otto von, 32
Bissau, 45

Boer War, 48
Bolama Bay, 29
Bolama Island, 45
Bosnia, peace missions, 188, 189, 191
Botelho Moniz coup (1961), 122, 124, 131, 140
Bowles, Charles, 133
Brazil
  Court transfer, 2, 4, 5, 7
  independence (1822), 10
  Indian occupation of Goa, Daman and Diu, 130
  official recognition of First Portuguese Republic (1910), 50
  Portuguese acknowledgement of independence (1825), 22
  Portuguese colonial policies, 133, 135
  slave trade, 24–5
Brazza, Pierre Savorgnan de, 28
Brioni Agreements (1991), 189
Britain
  African colonial disputes with Portugal, 29, 33–5
  African imperial project, 24, 29
  Anglo-German agreement (1898), 47–9, 54, 74
  Anglo-German agreement (1912–13), 52, 53, 54, 74
  Anglo-German rivalry, 47, 51, 52–3
  Anglo-Portuguese treaties, 31–2
  Azores military bases, 99–100, 104, 105, 106, 108
  Azores strategic value, 51
  balance of power concept, 3
  Cape to Cairo connection, 28, 29, 33–4
  Charles et George French ship issue, 25
  colonial investments, 27
  Continental Blockade, 2
  diplomatic rapprochement with Spain, 49, 52, 56, 74
  dynastic legitimacy concept, 3
  effective occupation principle, 33
  Entente Cordiale, 50
  European theatre in WWI, 68, 69, 70, 71–2
  Haldane Mission, 52–3
  Indian occupation of Goa, Daman and Diu, 130
  influence in Portugal, 3–4
  military aid to Portugal, 91
  NATO founding member, 112
  official recognition of First Portuguese Republic (1911), 50–1
  Peninsular Wars, 1
  Portugal as NATO founding member, 115
  Portuguese African colonies, 54–5
  Portuguese civil war (1846–1847), 20–1
  Portuguese civil wars (1820–1834), 4, 13
  Portuguese colonial policies, 133, 135, 136
  Portuguese defence plan (1940–1941), 106
  Portuguese defence plan (1943), 106–7, 109
  Portuguese slave trading, 23, 31
  post-WWII decline, 110
  proposed loan to Portugal, 48
  Quadruple Alliance (1815), 3, 4
  Quadruple Alliance (1834), 4, 15, 20
  rapid reaction forces, 170
  rearmament of Portugal, 104
  relations with Francoist Spain, 92–3
  Saint Malo Summit (1998), 170
  Second World War, 95, 96, 97
  secret Treaty with Portugal (1807), 2, 4
  Setembrismo, 16–17
  Spanish Civil War policy, 81
  trade agreement with Portugal (1810), 2, 4
  Treaty of Windsor (1899), 49
  "wolfram question", 101, 105
  see also Anglo-Portuguese Alliance
British Foreign Office, 52, 81
British South African Company, 40
Brussels Geographic Conference (1876), 29, 30
Brussels Pact countries, 112, 113
Buddhas, 79, 83
Buta, Álvaro, 44

Cabinda enclave, 42, 75, 147–8, 151
Cabora Bassa *see* Cahora Bassa
Cabral, Amílcar, 138, 146, 147, 148, 152, 154
Cabral, António Bernardo da Costa, 19–20, 21
Cabral, Vasco, 146, 147
caçadores, 7, 12, 14, 18, 21, 37, 86, 87
Cacheu, 45

Caetano, Marcello, 125, 137, 154, 158, 160
Cahora Bassa, 156, 157–8
Caldeira, Fernando Soares, 16
Canada, NATO founding member, 112
Canary Islands, 95, 176–7
Cândido Sarmento, Simão, 43
Canning, George, 12
Cape Verde Islands
African elites, 146
Anti-Colonial Movement (MAC), 148
independence movement leaders, 146, 147
Portuguese colonial empire, 22, 38–9
Portuguese territorial commands, 142
post-colonial technical-military cooperation, 180
Second World War, 95, 107
see also African Party for the Independence of Guinea and Cape Verde (PAIGC)
Capelo, Hermenegildo, 28, 30
Cardoso, António Maria, 33
Cardoso, Augusto, 30
Carlos de Borbón, 14–15
Carlos I, King of Portugal, 49
Carlota Joaquina, Queen of Portugal, 10
Cartagena, Pact of (1907), 49, 56, 74
Carvalho, Henrique de, 30, 43
Carvalho, Henrique Martins de, 113–14
Castro, Joaquim Pimenta de, 53, 59
CAUNC (Action Committee of the Cabinda National Union), 148
Central Permanent Commission of Geography, 30
CFSP (Common Foreign and Security Policy), 168, 169, 170, 175
Chagas, Pinheiro, 36
Chamber of Deputies, 11
Chamber of Peers, 9–10, 11, 17
Chamber of Senators, 17
Charles et George French ship, 25
China, 163
Chipenda, Daniel, 151
civil wars (1820–1834), 4, 8–15
CJTF (Combined Joined Task Forces), 170
CLSTP (Committee for the Liberation of São Tomé e Príncipe), 148, 149
Cold War
African continent, 137
Azores strategic value, 108
early inklings of, 109
end of, 167, 172
outbreak of, 111, 112
Combined Joined Task Forces (CJTF), 170
Committee for the Liberation of São Tomé e Príncipe (CLSTP), 148, 149
Common Foreign and Security Policy (CFSP), 168, 169, 170, 175
Common Security and Defence Policy (CSDP), 172, 175
Community of Portuguese-Speaking Countries (CPLP), 166
Concha, General, 20
CONCP (Conference of Nationalist Organisations of the Portuguese Colonies), 149
Congo, peace missions, 186
Congo estuary, 24
Congo Free State
Belgian imperial project, 29, 32, 43
demarcation of borders treaty (1891), 43
Congo treaty (1884), 31–2
Congo uprising (1913), 42, 44
Congo-Brazzaville, 151
Constitution (1822), 9–10, 11, 15, 16
Constitution (1838), 17
Constitutional Charter, 11–12, 15, 16, 17, 19, 20
Constitutional Governments (post-1976)
Angola's peace process, 186
Armed Forces, 166–7, 181–93
Atlanticism, 164–5, 166–8, 169, 170, 172–9
civil-military relations, 183–4
defence policy, 166–72, 183, 185, 190, 192–3
EEC membership (1986), 166, 168, 169, 173
Europeanisation, 165–72, 190
foreign policy, 164–72, 173, 185, 186, 190, 192–3
humanitarian assistance operations, 167, 173, 178, 181, 185, 192–3
military equipment modernisation programmes, 182–3

Constitutional Governments (post-1976) *(continued)*
peacekeeping operations, 181, 184–93
post-colonial relations, 166–7, 179–81
rapid reaction forces, 170
relations with NATO, 165, 167, 172–9
relations with US, 164–5, 167, 172, 173
Socialist Party governments (1996–2000), 169
Trade Agreement with EEC (1972), 165
Western European Union (WEU) membership, 169, 173
Constitutional Monarchy (1834–1910)
Armed Forces reforms, 18, 21, 35–8
foreign policy, 50
Mozambique military campaigns, 42
"pacification" campaigns, 39
replaced by First Republic, 50
threats to African empire, 47
Cordeiro, Luciano, 29, 30
Cordon, Vítor, 33
COREMO (Mozambique Revolutionary Committee), 155
Correia, Magalhães, 87
Cortes
African imperial project, 22–3
Army reform, 9
Constitution (1838), 17
Constitution approved (1822), 9–10
Miguel proclaimed absolute king (1828), 12
military recruitment, 9
Navy reform, 9, 14
Queen Carlota Joaquina's exile, 10
Corvo, Andrade, 30, 31, 32, 36
Costa, Afonso, 75, 76
Council of the Admiralty, 7, 9
Council of Europe, 165
Court, departure to Brazil, 2, 4, 5, 7
Coutinho, João de Azevedo, 34, 41
CPLP (Community of Portuguese-Speaking Countries), 166
Criminal Investigation Police, 87
Croatia, NATO membership, 174–5
Cruz, Viriato da, 147, 148
CSDP (Common Security and Defence Policy), 172, 175
Cuamato, 42, 43
Cuamato Grande, 43
Cuamato Pequeno, 43
Cuanhama, 43
Cutileiro, José, 169
Cyprus, peace missions, 186
Czech Republic, NATO membership, 174

Dahomey, 130
Daman and Diu, 39, 129–30
Dayton Agreements, 188
decolonisation
Armed Forces Movement (AFM) (1974), 162
defined, 127–8, 146
pre-constitutional government, 162–3
Salazar's opposition to, 100, 111
see also liberation movements
decolonisation wars
Air Force, 142–3, 144–5, 154
Algeria, 136, 139, 145, 160
Angola, 130–1, 141, 142, 143, 150–1, 159
Armed Forces, 140–3, 144–5, 150, 151, 153, 154, 155–7, 158, 159–60
Army, 140, 141, 142, 144, 158, 159–60
Estado Novo, 130–1, 139–46
Guinea-Bissau, 126, 131, 138, 141, 142, 152–4, 159, 160
Mozambique, 131, 155–9, 160
Navy, 142–3, 144
Delgado, Humberto, 117
Dembos, 42
democratisation
Armed Forces Movement (AFM) (1974), 162
Armed Forces role, 181, 183
Portugal's admission to NATO, 117
post-WWII growth of, 109, 110
pre-constitutional government, 162, 163
Salazar's opposition to, 100
Denmark, rearmament of Portugal, 104
Deslandes, Venâncio, 150
Dili, 45
diseases, 26
Dominican Republic, peace missions, 186
Duroselle, Jean-Baptiste, 114
dynastic legitimacy, 2, 3–4, 9

East Timor
peace missions, 190–1
see also Portuguese Timor
Eça, António Júlio da Costa Pereira de, 66
Edward VII, King of Great Britain, 49
effective occupation principle, 29–30, 32–3
Egypt, 29
elites, 146, 147, 155
Elizabeth II, Queen of Britain, 129
Enes, António, 40
England *see* Britain
English Alliance *see* Anglo-Portuguese Alliance
ESDI (European Security and Defence Identity), 168–9, 170, 173, 175
ESDP (European Security and Defence Policy), 170–2, 175
Estado Novo (1933–1974)
Air Force rearmament, 104, 121
Air Force reforms, 143, 144–5
Anglo-Portuguese alliance, 77, 81
Armed Forces rearmament, 90–1, 103–5, 121, 126, 136, 137–8, 143–5
Armed Forces reforms, 89–91, 103, 104–5, 119–22, 140–2, 143, 144–5
Army rearmament, 90–1, 103–4, 121, 143, 144
Army reforms, 89–91, 103, 120, 142
Botelho Moniz coup (1961), 122, 124, 131, 140
British military aid, 91
civil-military relations, 89, 105–6, 116, 122, 160–1
colonial policy, 117, 123, 127–8, 129–37, 140
"cooperative neutrality", 98, 99, 100, 101, 104, 106
decolonisation wars, 130–1, 139–46
defence agreement with US (1957), 118, 123–4
defence agreement with US (1971), 125–6
defence policy (pre-1939), 88–91
defence policy (1939–1945), 103–9
defence policy (1950s), 117–22
defence policy (1961–1974), 125, 136, 139–40
establishment of, 77
foreign policy (pre-1939), 77–80, 92–4
foreign policy (1945--1949), 94–102, 110–12
foreign policy (1950s), 117–18, 122, 128–30
foreign policy (1960s), 123–4, 129, 130–4, 136
foreign policy (1970s), 125–6, 129
"geometric neutrality", 97, 98, 99, 101, 106
international isolation of (1961–1974), 128–39
international marginalisation (1945–1949), 110–12
Lajes Agreement (1948), 110, 111
Mealhada uprising (1947), 105, 122
national military industry, 144
NATO founding member, 112–17, 128
NATO reduced commitments, 124–5, 136, 139–40
Navy rearmament, 91, 104–5, 121
Navy reforms, 91, 103, 104–5, 120–1, 143, 144
Nazi Germany threat, 103, 106
opposition movements, 78–9
"peninsular friendship" notion, 77–8, 84, 94, 130
political consolidation, 85, 90, 93
post-WWII "communist danger", 109
relations with France, 134, 136, 144, 145
relations with Spain, 78–80, 84, 98, 109
relations with US, 100–1, 111, 117–18, 121, 122, 123–4, 125–6, 128, 130, 132–4, 136, 137–8
relations with West Germany, 134, 136, 144, 145
reviralho, 78–9, 80
Second World War goals/objectives, 93–5, 96, 99, 100, 101–2
Spanish Civil War policy, 81–5, 88
Spanish Republic threat, 78–9, 80, 88
state visits, 129
United Nations admission (1955), 117, 123, 128
US military cooperation agreement (1948), 111

Estado Novo (1933–1974) *(continued)*
US mutual assistance agreement (1951), 118, 121
"wolfram question", 101, 105
Esteves, Raul, 83, 118, 120
ethno-nationalist movements, 147–8
Euro-Atlantic Partnership Council, 174
Eurocorps, 169, 170
Eurofor (European Rapid Operational Force), 169–70
Euromarfor (European Maritime Force), 170
Europe
colonial investments, 26–7
demographic expansion (1800s), 26
Marshall Plan, 111
scramble for Africa, 26–35
technological revolution, 26
European Community (EEC), 165, 166, 168, 169, 173
European Defence Community, 168
European Maritime Force (Euromarfor), 170
European Political Cooperation, 168
European Rapid Operational Force (Eurofor), 169–70
European Security and Defence Identity (ESDI), 168–9, 170, 173, 175
European Security and Defence Policy (ESDP), 170–2, 175
European theatre, First World War, 54, 57, 60, 65, 66–72, 74
European Union
Balkans peace missions, 188–90, 191
Horn of Africa peace missions, 192
Maastricht Treaty on European Union (1991), 168
military dimension, 170–2
Portuguese membership, 167
Portuguese presidency (1992), 169
relations with NATO, 175–6, 179
WEU incorporation, 170–2
Évoramonte Convention (1834), 15

Falange, 83, 84
Ferdinand VII, King of Spain, 10, 14
Ferreira, José Medeiros, 114
Ferry, Jules, 32
First Portuguese Republic *see* Republic (1910–1926)
First World War
African theatre, 41, 42, 43, 54, 60, 65–6, 67, 74
Anglo-Portuguese Alliance, 56, 57, 73–4
Angola, 43, 54, 65–6, 74
colonial thesis, 54–5
European theatre, 54, 57, 60, 65, 66–72, 74
European-peninsular thesis, 54, 55–8
internal politics thesis, 54, 58–60
Mozambique, 41, 42, 54, 65, 66, 74
Portuguese Air Force, 86
Portuguese war goals/objectives, 53–60, 65, 72–6
roulement problem, 68, 70, 71
trench warfare, 61, 66–7, 69–70
Zambezia uprising, 41
FLAD (Luso-American Development Foundation), 165
FLEC (Front for the Liberation of the Enclave of Cabinda), 148
FNLA (National Liberation Front of Angola), 147, 150–1
Fonseca Magalhães, Rodrigo da, 19
Fontes Pereira de Melo, António Maria de, 35, 36
FRAIN (African Revolutionary Front for the National Independence of the Portuguese Colonies), 148–9
France
African imperial project, 24, 29
Algeria colonisation, 24, 29
Algeria decolonisation war, 136, 139, 145, 160
Charles et George French ship issue, Mozambique, 25
colonial investments, 27
Congo treaty (1884), 32
Continental Blockade, 2
Convention with Portugal (1886), 33
Dakar to Djibouti connection, 28
decolonisation, 130
Entente Cordiale, 50
Eurocorps, 169
European theatre in WWI, 68
Franco-German agreement (1911), 51
Franco-Spanish agreement (1912), 51
invasion of Spain (1823), 3–4, 10
invasions of Portugal, 1
NATO founding member, 112
NATO return (2009), 175
NATO withdrawal (1966), 124, 175

official recognition of First Portuguese Republic (1911), 50
Peninsular Wars, 1
Portuguese African colonies, 54
Portuguese Armed Forces rearmament, 136, 144, 145
Portuguese civil war (1846–1847), 20
Portuguese civil wars (1820–1834), 4, 13
Portuguese colonial policies, 135, 136
Portuguese–French Convention (1896), 45
Quadruple Alliance (1834), 15, 20
rapid reaction forces, 170
relations with Portugal, 134, 136, 144, 145
Revolution (1830), 16
Saint Malo Summit (1998), 170
Setembrismo, 16
Spanish Civil War policy, 81
Franco, Francisco
diplomatic negotiations with Hitler, 97
Estado Novo support for, 80, 81
meeting with Salazar (1942), 98
relationship with Salazar, 94
Salazar's support for, 80
Second World War, 96
Spanish Civil War victory, 84
Franco-Prussian War (1871), 27
Freire de Andrade, Gomes, 5, 7, 40
FRELIMO (Mozambique Liberation Front), 131, 148, 149, 155, 156, 157, 158–9
French Revolution, influence in Portugal, 1
Front for the Liberation of the Enclave of Cabinda (FLEC), 148
fula ethnic group, 153

Galvão, Henrique, 131, 150
Gamito, Pedro, 28
Gaza campaign, 39, 40–1
German South West Africa, 43
Germany
African imperial project, 29
African theatre in WWI, 43, 54, 65
Anglo-German agreement (1898), 47–9, 54, 74
Anglo-German agreement (1912–1913), 52, 53, 54, 74
Anglo-German rivalry, 47, 51, 52–3
claim to Kionga, 75
colonial empire loss, 93
colonial investments, 27
Congo treaty (1884), 32
Convention with Portugal (1886), 33
emergence as international power, 27, 34, 47
Eurocorps, 169
European theatre in WWI, 67, 70, 71
Franco-German agreement (1911), 51
Franco-Prussian War (1871), 27
Haldane Mission, 52–3
invasion of Poland (1939), 92, 94
negotiations with Spain during WWII, 96, 97
Operation Barbarossa, 97
Operation Felix, 97, 106
Portuguese diplomatic approach, 34
post-WWII division, 112
proposed loan to Portugal, 48
rearmament of Portugal (WWII), 103
reunification (1990), 167
Second World War, 95, 96, 97, 98, 99
territorial claims in Portuguese Africa, 55
as threat to Portugal, 103, 106
unification (1871), 27
see also Prussia; West Germany
Gibraltar, 3, 93, 95, 96, 97
Goa, 39, 129–30
Golan Heights, peace missions, 186
Gomes, Costa, 151, 160
Gomes, Henrique de Barros, 33, 34
Gonçalves, Vasco, 162
GRAE (Revolutionary Government of Angola in Exile), 147, 148
Gramido, Convention of, 21
Grant, Ulysses S., 29
Grumete ethnic group, 45
guerrilla warfare
Absolute Monarchy, 6, 10
Angola decolonisation war, 150–1
Constitutional Monarchy, 17, 20
counter-guerrilla techniques, 137–8, 139, 140, 142, 145–6, 160
decolonisation wars, 131–2, 139, 150–1, 152–4, 155–8

guerrilla warfare *(continued)*
  Guinea-Bissau decolonisation war, 152–4
  Mozambique decolonisation war, 155–8
  Sakwa Plan, 131–2
  stages of, 152
Guinea-Bissau
  African elites, 146
  Africanisation of Portuguese military, 141, 160
  Anti-Colonial Movement (MAC), 148
  decolonisation wars, 126, 131, 138, 141, 142, 152–4, 159, 160
  ethnic diversity, 45
  FRAIN membership, 149
  Independence declaration (1973), 139, 154, 163
  liberation movement leaders, 147
  liberation movements, 139, 147, 149, 152, 153–4, 159
  Operation Green Sea, 153
  "pacification" campaigns, 39, 45, 64
  Portuguese independence recognition (1974), 163
  Portuguese special troops, 142
  Portuguese territorial commands, 142
  Portuguese territorial occupation, 38–9
  post-colonial technical-military cooperation, 180
  strike by dockers (1959), 149
Guinea-Conakry, 147, 152, 153
Guname, Paulo, 155
Gungunhana, Emperor of Gaza, 39, 40–1
Gwambe, Adelino, 155

Haile Selassie, Emperor of Ethiopia, 129
Haldane Mission, 52–3
Hammarskjöld, Dag, 129
Harriman, Averell, 134
Hastings, Adrian, 139, 158
Helsinki European Council, 171
Hezbollah, Lebanon War (2006), 192
historic rights principle, 28–9, 30, 33
Hitler, Adolf
  diplomatic negotiations with Franco, 97
  Operation Felix, 97
Holy Alliance, 2–3
Holy See, normalisation of relations with Portugal (1848), 4
Home for the Students of the Empire, 146
Huambo, 44
Huíla Plateau, 43
humanitarian assistance operations, 167, 173, 178, 181, 185, 192–3
Humbe, 66
Hungary, NATO membership, 174

Iberian Bloc, 98, 109, 115, 118, 120
Iberian Pact, 84, 94, 96, 99–100, 115
IBERLAND, 124, 165, 176
IFOR (Implementation Force), 188, 189
independence movements *see* liberation movements
Independent Mixed Brigade (BAI), 165
India
  Anglo-Portuguese treaties, 31, 32
  British colonisation, 24
  "pacification" campaigns, 64
  peace missions, 186
  Portuguese imperialism, 39, 123, 129–30
Indian Union, 129–30
Indonesia, East Timor referendum, 190
Injai, Abdul, 45
Institute of High Military Studies, 103
International Association of the Congo, 29, 32
Iran
  Islamic Revolution (1979), 164
  peace missions, 186
Iraq
  peace missions, 186
  US invasion (2003), 175
Isabel II, Queen of Spain, 14
ISAF (International Security Assistance Force), 191–2
Israel
  Lebanon War (2006), 192
  Yom Kippur War, 138
Italy
  Portuguese African colonies, 54
  rapid reaction forces, 170
  rearmament of Portugal, 103–4
  unification (1871), 27
Ivens, Roberto, 28, 30

Jardim, Jorge, 158

João VI, King of Portugal, 2, 11
Jordana, Count, 98
José, Amaro, 28
Junot, Jean Andoche, duc d'Abrantès, 5

Kampala Episcopal Conference, 138
Katrina, Hurricane, relief operations, 177
Kawandame, Lázaro, 155
KFOR (Kosovo Force), 189
Kionga, 75
Kissinger, Henry, 126, 137, 138
Kosovo, peace missions, 188–90
Kosovo crisis, 171, 178
Krupp's, 63
Kubichek, Jucelino, 129

Lafões, João Carlos de Bragança, Duke of, 5
Lajes Agreement (1948), 110, 111
Lajes Agreement (1957), 118, 123, 132
Lajes Agreement (1971), 138
Lajes Agreement (1979), 164
Lajes Agreement (1983), 164
Lara, Lúcio, 146, 148
League of Nations (LN), 74, 76, 77, 79, 82, 110
Lebanon, peacekeeping operations, 186, 190, 192
Legião Portuguesa ("Portuguese Legion"), 5, 90
Léopold II, King of the Belgians, 30, 32, 43
Liberal Revolution (1820), 8–9
  international context, 1–5
  peninsular context, 4–5
  Spanish support, 4–5
liberal wars (1820–1834), 4, 8–15
Liberalism
  different views on, 15
  establishment of, 1, 3, 15
liberation movements, 146–50
  Anderson plan, 134
  Angola, 131, 147–8, 149, 150–1, 159, 180
  common denominators among, 148, 159
  communist influence, 137
  Guinea-Bissau, 139, 147, 149, 152, 153–4, 159
  leadership, 146–7
  Mozambique, 131, 148, 149, 155, 156, 157, 158–9
  political confrontation, 123, 129
  Portuguese military strategy, 160
  transfer of power to, 163
  US ends contact with, 137
  US support for, 131
  Vatican support for, 133, 138
Lippe, Wilhelm Graf zu, 5
Lisbon Geographic Society, 30, 31, 33
Lisbon, Treaty of (2010), 171, 172
Livingstone, David, 28, 30
London Committee, 81, 82
Louis Philippe, King of France, 4, 16
Lourenço Marques
  Anglo-Portuguese dispute (1875), 29
  British customs exemptions, 31
  construction of the port, 31
  "pacification" campaigns, 40
  Portuguese Air region centre, 143
  strategic importance of, 48
  strike by dockers (1956), 149
Lourenço Marques Treaty (1879), 32
Luanda
  independence movements, 147
  Portuguese Air region centre, 143
  Portuguese "public work expeditions", 30
  prison assault (1961), 150
Luís do Rego, 10
Lundas, 43
Lusaka Protocol (1994), 187
Luso-American Development Foundation (FLAD), 165
Luxembourg
  Eurocorps, 169
  NATO founding member, 112
Lys, Battle of the, 69, 70, 71

MAC (Anti Colonial Movement), 148
Macau, 39, 142, 163
MacDonnell, Ranald, 20
Machel, Samora, 156
MacMahon, Edme Patrice Maurice, comte de, 29
MacMillan, Harold, 130
Macololos, 33, 34
Macua zone, 41
Madeira
  Liberal Wars (1828–1834), 13
  military divisions, 18, 21, 36, 62, 86
  uprising (1931), 87
  WWII mobilisation, 107
Magul, Battle of (1895), 40
Majanga da Costa, 41

Makonde ethnic group, 155
Makonde Plateau, 155, 156, 157
Makua ethnic group, 155
Malanje, 43
malaria prophylaxis, 26
Malawi, 148, 155
Malawi Lake, 28
Mandinga ethnic group, 45
Manjaco ethnic group, 45
MANU (Mozambique African National Union), 148
Maquela do Zombo, 42
Maria Christina, Queen of Spain, 4
Maria da Fonte Revolt (1846), 19–20
Maria II, Queen of Portugal, 11, 12, 15, 16–17, 20, 21
Maria Pia Fortress, 43
Marracuene, Battle of (1895), 40
Marshall Plan, 111
Martin, George, 7
Martinhada, 8
Mataca, chief of the Ajaua, 42
Matos, Norton de, 67
Maúta, 66
Maziua, 65
Mealhada uprising (1947), 105, 122
Medeiros, Aurélio José de, 8
Mercantile Union Company, 25
Merida, 81–2
Meridional Africa, 28, 33–4
Metternich, Clemens Wenzel Lothar, Fürst von, 12
MFA (Armed Forces Movement) (1974), 141, 160–1, 162, 183
MIA (Movement for the Independence of Angola), 147
Miguel I, King of Portugal, 10–11, 12–13, 14–15
Miguelism, 9
Military College, 18
military dictatorship (1926–1933), 77, 85–8, 90
military industries innovation, 26
militias
  Cortes decree (1821), 9
  English reorganisation of, 5, 6, 7
  First Republic reforms (1911), 61, 64
  Junot's decrees, 5
  Liberal Wars (1828–1834), 12, 14
  Vilafrancada regime, 10
MING (Movement for the National Independence of Guinea), 147
Minho, Maria da Fonte Revolt (1846), 19–20
MLEC (Movement for the Liberation of the Enclave of Cabinda), 147–8
Mocidade Portuguesa ("Portuguese Youth"), 90, 146
Mondlane, Eduardo, 131, 146, 148, 156
Mongua, Battle of, 66
Moniz, Egas, 75, 76
Moniz, Jorge Botelho, 83
Moniz, Júlio Botelho, 122, 124, 131, 139, 140
Monroe doctrine, 50
Monteiro, Correia, 28
Mormugao Port, 31
Movement for the Independence of Angola (MIA), 147
Movement for the Liberation of the Enclave of Cabinda (MLEC), 147–8
Movement for the National Independence of Guinea (MING), 147
Mozambique
  African elites, 146
  Africanisation of Portuguese military, 141, 158, 160
  Anglo-German agreement (1898), 48, 74
  Anglo-Portuguese disputes, 29
  Anti-Colonial Movement (MAC), 148
  Cahora Bassa, 156
  Charles et George French ship issue, 25
  colonial army, 37–8
  decolonisation wars, 131, 155–9, 160
  dockers' strike (1956), 149
  ethno-nationalist movements, 148
  First World War, 41, 42, 54, 65, 66, 74
  Gaza "pacification" campaign, 39, 40–1
  German claim to Kionga, 75
  German territorial claims, 55
  independence movement leaders, 146–7
  liberation movements, 131, 148, 149, 155, 156, 157, 158–9
  Meridional Africa, 28, 33–4

Nyassa "pacification" campaign, 41–2
Operation Frontier, 157
Operation Gordian Knot, 156, 157
Organic Charter, 37
"pacification" campaigns, 39, 40–2, 45, 64
peace missions, 186, 187
Portuguese Air regions, 143
Portuguese early expeditions, 28
Portuguese interior penetration zones, 22
Portuguese naval command, 143
Portuguese "public work expeditions", 30–1
Portuguese special troops, 142
Portuguese territorial commands, 142
Portuguese territorial occupation, 23, 24, 33, 38–9, 40–2
post-colonial technical-military cooperation, 180
South African claims, 75
uprising against compulsory cotton production, 149
Versailles Peace Conference (1919), 75
western border defined (1891), 34
Wiriamu massacre, 139, 158
WWII mobilisation, 107
Zambezia "pacification" campaign, 41
see also Lourenço Marques; Nyassa; Tete
Mozambique African National Union (MANU), 148
Mozambique Company, 41
Mozambique Institute, 155
Mozambique Liberation Front (FRELIMO), 131, 148, 149, 155, 156, 157, 158–9
Mozambique Revolutionary Committee (COREMO), 155
MPLA (People's Movement for the Liberation of Angola), 147, 149, 150, 151
MUD-Juvenil, 146
Munich Conference (1938), 92
Munich crisis (1938), 84, 93–4, 103
Municipal Guard, 62
Mutu-a-Quebera, 44

Namarrais ethnic group, 41
Namibia, peace missions, 186
Napier, Charles, 14
Napoleon I, Emperor of the French, 2
National African Union of Independent Mozambique (UNAMI), 148
National Democratic Movement of Mozambique (UDENAMO), 148, 149
National Guard, 16–17
National Liberation Front of Angola (FNLA), 147, 150–1
National Union for the Total Independence of Angola (UNITA), 148, 151
National Union of Workers of Guinea-Bissau, 147
NATO *see* North Atlantic Treaty Organisation (NATO)
Naulila, Battle of, 65
Navy
Absolute Monarchy reforms, 7–8
Admiralty Council, 7, 9
cadre of officers, 7
Constitutional Monarchy reforms, 36
decolonisation wars, 142–3, 144
defence plan (1943), 107
Estado Novo rearmament plans, 91, 104–5, 121
Estado Novo reforms, 91, 103, 104–5, 120–1, 143, 144
First Republic reforms, 64
IBERLAND, 124, 165
Liberal Wars (1828–1834), 13–14
liberalist period reforms, 15–16
Miguelista government, 12–13
military dictatorship reforms, 86, 87–8
military expenditure, 88, 91, 105
NATO military commitments, 119, 124, 165
numbers of personnel, 9
recruitment system, 9, 36
Royal Navy Academy, 7
Royal Navy Brigade, 7
"sailors' revolt" (1936), 87–8
Second World War, 104–5, 107–8
transfer to the Azores, 107–8
Vintista reforms, 9
Navy Ministry, 119
Negomano, 66
Nehru, Jawaharlal, 130
Nekaka, Barros, 147

Netherlands
NATO founding member, 112
rapid reaction forces, 170
Neto, Agostinho, 138, 146, 148, 151
Nevala, 66
New Guinea, peace missions, 186
Nhamacurra, 66
Nice, Treaty of (2000), 170, 171, 175
Nkrumá, Kwame, 148
Nogueira, Albano, 114
Nogueira, Franco, 114, 132, 134
Normandy, 99, 100
North Atlantic Cooperation Council, 174
North Atlantic Treaty Organisation (NATO)
Afghanistan peace mission, 191–2
Azores strategic value, 113
Balkans peace missions, 188–90, 191
Berlin Summit (1996), 175
Brussels Summit (1984), 174
Constitutional Governments, 165, 167, 172–9
eastern enlargement, 174–5, 178
Estado Novo, 112–17, 124–5, 128, 136, 139–40
founding of (1949), 112
French return (2009), 175
French withdrawal (1966), 124
humanitarian assistance operations, 178
IBERLAND, 124, 165, 176
influence on Portuguese defence policy, 117–22
Lisbon Summit (2010), 175, 178–9
Madrid Summit (1997), 174
Partnership for Peace, 174
peacekeeping operations, 178, 188–90, 191–2
Portugal as founding member, 104, 106, 112–17, 128
Portuguese commitments, 119, 124–5, 136, 139–40, 165, 173
Portuguese Navy rearmament, 105
post-Cold War restructuring of military commands, 176–7
post-Cold War strategic concepts, 177–9
post-Cold War transformation, 173–9
pre-constitutional government, 164
rapid reinforcement of Europe doctrine, 100, 113, 132
relations with European Union, 175–6, 179
relations with Western European Union (WEU), 168
Rome Summit (1991), 174, 177–8
SACEUR, 176, 177
SACLAND, 124, 176–7
Spain's exclusion, 114, 115, 116
Spanish membership, 176, 177
transatlantic relations, 175
Washington Summit (1999), 174, 177, 178
Nyassa
Mozambique decolonisation war, 155, 156
"pacification" campaigns, 41–2
Portuguese "public work expeditions", 30
Portuguese territorial claim, 31
Portuguese territorial occupation, 33
Nyassa Chartered Company, 41, 42
Nyerere, Julius, 148

OAU (Organisation of African Unity), 135–6, 151
Old Regime *see* Absolute Monarchy (pre-1834)
ordenanças
Cortes decree (1821), 9
English reorganisation of, 5, 6–7
Liberal Wars (1828–1834), 12, 14
Vilafrancada regime, 10
Organisation of African Unity (OAU), 135–6, 151
Ornelas, Mouzinho de Albuquerque Ayres d', 38, 40, 41
Ovimbundu ethnic group, 42, 44, 148

PAI (African Party for the Independence), 147
PAIGC *see* African Party for the Independence of Guinea and Cape Verde (PAIGC)
Pais, Sidónio, 54, 70, 71
Paiva, Artur, 33, 43, 44
Paiva Couceiro, Henrique Mitchell de, 33, 38, 40, 43, 52, 61
Paiva de Andrade, Joaquim Carlos, 33
Pakistan
earthquake relief operations, 177
peace missions, 186

Palmela, Pedro Sousa Holstein, Duke of, 13, 15, 20
Palmerston, Henry John Temple, Viscount, 13, 23
PALOP (Portuguese Speaking African countries), 166, 180, 181, 184
Panama Canal Zone, 145
Papéi ethnic group, 45
Paris Founding Act (1997), 174
Party of the United Struggle for Africans in Angola (PLUAA), 147
Passos Manuel, 16, 17
Patuleia civil war, 20–1
Paul VI, Pope, 138
PCA (Angolan Communist Party), 147
PCP (Portuguese Communist Party), 147, 164
Pedro I, Emperor of Brazil (Pedro IV of Portugal), 4, 11, 12, 13–14, 15
Pedro II, Emperor of Brazil, 13
Peninsular Pact, 84, 94, 96, 99–100, 115
"peninsular political system", 4–5, 14
Peninsular Wars, 1, 4
People's Movement for the Liberation of Angola (MPLA), 147, 149, 150, 151
Pereira, Aristides, 154
Pereira, Teotónio, 84
Petersberg tasks, 170, 171
PIDE (*Polícia Internacional e de Defesa do Estado*), 142
Pink Map, 33, 34
PLUAA (Party of the United Struggle for Africans in Angola), 147
Poland
  German invasion (1939), 92, 94
  NATO membership, 174
police, 7, 62, 87, 90
Polícia de Vigilância e Defesa do Estado (PVDE), 90
Polícia Internacional e de Defesa do Estado (PIDE), 142
Polytechnic School, 18
Portugal *see* Absolute Monarchy (pre-1834); Constitutional Governments (post-1976); Constitutional Monarchy (1834–1910); *Estado Novo* (1933–1974); military dictatorship (1926–1933); pre-constitutional government (1974–1976); Republic (1910–1926)
Portuguese Armed Forces
  Absolute Monarchy reforms, 5–8, 9, 11–12
  Africanisation of troops, 141, 142, 158, 160
  Angola decolonisation war, 140, 141, 142–3, 144–5, 150, 151, 154, 158, 159–60
  autonomy of, 7
  civil-military relations, 64, 89, 105–6, 116, 122, 160–1, 183–4
  compulsory military service, 6, 35, 62, 63, 67, 85–6, 89, 140–1, 181–2
  Constitutional Government reforms, 166–7, 181–93
  Constitutional Monarchy reforms, 18, 21, 35–8
  counter-guerrilla warfare, 137–8, 139, 140, 142, 145–6, 160
  decolonisation wars, 140–3, 144–5, 150, 151, 153, 154, 155–7, 158, 159–60
  defence plan (1940–1941), 106, 108–9
  defence plan (1943), 106–7, 108–9
  democratic model, 181, 183–4
  democratisation role, 181, 183
  English reorganisation of, 5–7
  Estado Novo rearmament, 90–1, 103–5, 121, 126, 136, 137–8, 143–5
  Estado Novo reforms, 89–91, 103, 104–5, 119–22, 140–2, 143, 144–5
  First Republic reforms, 61–4
  Freire de Andrade's reforms, 5
  Guinea Bissau decolonisation war, 141, 142, 153, 154
  humanitarian assistance missions, 167, 181, 185, 192–3
  inclusion of women, 181
  length of military service, 6, 9, 86, 89, 141, 181
  liberalist period reforms, 15–16, 17–19, 21
  Lippe's reforms, 5
  military dictatorship reforms, 85–8, 90
  military divisions, 18, 21, 35, 36, 62, 86
  military expenditure, 88, 91, 105, 122, 145

Portuguese Armed Forces *(continued)*
military state of (pre-WWI), 61
Minister of National Defence post created (1950), 119
Ministry of National Defence (MDN) created (1982), 184
money remissions, 18, 35, 37, 63–4, 86, 89
Mozambique decolonisation war, 155–7, 158
national military industry, 144
number of personnel (1917), 68
number of personnel (1939–1943), 107
number of personnel (1961–1974), 140
number of personnel (1974–2010), 182
Operation Frontier, 157
Operation Gordian Knot, 156, 157
Operation Green Sea, 153
peacekeeping operations, 181, 184–93
Portugal as NATO founding member, 116
Portuguese membership of NATO, 118–19
post-colonial technical-military cooperation, 179–81
professionalisation, 7, 116, 122, 181–2
psychological action operations, 153, 156, 160
public trust in, 185
recruitment system, 6, 7, 9, 17–18, 35, 37, 62, 89, 140–1
Regeneration (1851) impact, 35
repetition schools, 62, 64
Second World War, 103–8
WWII mobilisation, 107
see also Air Force; Army; *caçadores*; First World War; militias; Navy; *ordenanças*
Portuguese civil war (1846–1847), 20–1
Portuguese civil wars (1820–1834), 4, 8–15
Portuguese Communist Party (PCP), 147, 164
Portuguese Democratic Party, 59–60, 67
"Portuguese Legion" (*Legião Portuguesa*), 5, 90
Portuguese Military Observation Mission, 82–3
Portuguese Overseas Council, 24
Portuguese Republic (1933–1974) *see Estado Novo* (1933–1974)
Portuguese Revolution (1910), 50, 61
Portuguese Speaking African countries (PALOP), 166, 180, 181, 184
Portuguese Timor
"pacification" campaigns, 39, 45–6, 64
Portuguese colonial empire, 39
Portuguese territorial commands, 142
Second World War, 101, 108
see also East Timor
Portuguese War Ministry, 119
"Portuguese Youth" (*Mocidade Portuguesa*), 90, 146
pre-constitutional government (1974–1976)
decolonisation, 162–3
democratisation, 162, 163
foreign policy, 162–4
Lajes Agreement, 138
NATO membership, 164
relations with the Soviet Union, 163
third-worldism, 163, 165
transition to democracy, 162, 163–4
Princess Amelia Fortress, 43
pronunciamentos, 8, 12, 17, 19
Prussia
Franco-Prussian War (1871), 27
Holy Alliance, 2–3
trade agreement with Portugal (1834), 4
see also Germany
PVDE (*Polícia de Vigilância e Defesa do Estado*), 90

Quadros, Jânio, 133
Quadruple Alliance (1815), 3, 4
Quadruple Alliance (1834), 4, 15, 16, 20
Quangar, 65
Quionga, 66

Ranque Franque, Luís, 148
Regeneration (1851), 18, 21, 24, 35
Renouvin, Pierre, 114
Republic (1910–1926)
Armed Forces reforms, 61–4
civil-military relations, 64

collapse of, 77
dominant party system, 59–60
First World War, 65–72
foreign policy, 50, 51
international recognition, 50–1, 57
international standing, 47–53, 74
Mozambique military campaigns, 42
national legitimisation, 58, 76
objectives and political strategies, 53–60
"pacification" campaigns, 39
political consolidation, 58, 60, 76, 93
relations with Spain, 56
threats to African empire, 47
Republican National Guard, 62, 64, 87
Requetés militias, 83
revolution (1974) *see* Armed Forces Movement (MFA) (1974)
Revolutionary Government of Angola in Exile (GRAE), 147, 148
Rhodes, Cecil, 29, 33–4, 40
Rhodesia, 136, 148
Roberto, Holden, 131, 147, 151
Roçadas, Jose Augusto Alves, 42, 43, 65, 66
Romania, NATO membership, 174
Rome Peace Accords (1992), 187
Roussillon Campaign (1793), 5
Royal Navy Academy, 7
Royal Navy Brigade, 7
Royalist forces, 12, 13, 14, 15, 20
Rusk, Dean, 123, 132
Russian Empire, Holy Alliance, 2–3
Russian Federation
Angola's peace process, 186
association to NATO, 174
see also Soviet Union

Sá da Bandeira, Marquis
African imperial project, 22–3, 24, 25, 31
Belémzada, 17
civil war (1846–1847), 20
government appointment, 16
military reforms, 17
Sá Nogueira, Bernardo, 13, 15
Sadowa, Battle of, 35
Saint Malo Summit (1998), 170
Sakwa Plan, 123, 131–2
Salazar, Antonio de Oliveira
Anderson Plan, 133, 134
Anglo-American military bases in the Azores, 101, 132, 133
Anglo-Portuguese alliance, 77, 94
anti-communism, 109, 114, 115, 159
colonial policy, 117, 123, 129–30, 132–3, 134, 136, 140, 163
defence agreement with US (1957), 123–4
defence policy, 88–9, 91
distrust of US, 100–1, 111, 115, 131
establishment of the *Estado Novo*, 77
foreign policy, 77–8, 79, 80, 94–5, 96, 97, 98, 101–2, 110–11
meeting with Franco (1942), 98
opposition to decolonisation, 100, 111
opposition to democratisation, 100
"peninsular friendship" notion, 84
Portugal as NATO founding member, 114–16
Portuguese "geometric neutrality", 97, 98
Portuguese neutrality in WWII, 94–5, 96, 97, 98, 101–2
relationship with Franco, 94
relationship with USA, 98–9, 100
retirement from office (1968), 125, 137, 160
Sakwa Plan, 132, 133
Spanish Civil War policy, 80, 81, 82, 84–5
Spanish Republic as "red threat", 78, 79, 80
support for Franco, 80
United Nations founding, 110–11
Saldanha, Marshall, 12, 15, 17, 20, 21
Salisbury, Robert Cecil, Marquess of, 30
San Francisco Conference (1945), 110
Santo, Guilherme Espírito, 148
Santos Costa, Fernando, 89, 90–1, 103, 109, 118, 120, 122
Santos, Machado, 61
Santos, Marcelino dos, 138, 146–7, 148, 156
São João Baptista de Ajudá, 39, 130
São Salvador, 42
São Tomé e Príncipe
Anti-Colonial Movement (MAC), 148
"monkey hunt" uprising (1953), 149

São Tomé e Príncipe *(continued)*
Portuguese colonial empire, 22, 38–9
Portuguese territorial commands, 142
post-colonial technical-military cooperation, 180, 181
Sartorius, George Rose, 13, 14
Savimbi, Jonas, 146, 148, 151
Schultz, Arnaldo, 152–3
Second Portuguese Republic *see Estado Novo* (1933–1974)
Second World War
Anglo-Portuguese Alliance, 93–4, 96
Azores, 95, 96, 106, 107–8
breakout of (1939), 92, 94
Britain, 95, 96, 97
Germany, 95, 96, 97, 98, 99
Nazi Germany threat to Portugal, 103, 106
Operation Barbarossa, 97
Operation Felix, 97, 106
Operation Torch, 98
Portuguese Armed Forces, 103–8
Portuguese "cooperative neutrality", 98, 99, 100, 101, 104, 106
Portuguese defence plan (1940–1941), 106, 108–9
Portuguese defence plan (1943), 106–7, 108–9
Portuguese defence policy, 103–9
Portuguese "geometric neutrality", 97, 98, 99, 101, 106
Portuguese neutrality, 94–102, 106
Portuguese war goals/objectives, 93–5, 96, 99, 100, 101–2
Soviet Union, 97, 98
Spain, 95–7, 98, 102
United States, 98–9
self-determination principle, 111, 127–8, 132, 162–3
Senegal, 147, 152
Senghor, Leopold, 152, 154
Sepúlveda, General, 10
Serpa Pinto, Alexandre Alberto da Rocha de, 30, 33, 34
Setembrismo, 16–19, 20, 23
SFOR (Stabilisation Force in Bosnia and Herzegovina), 189, 191
Shire Valley, 30, 34
Sicily, 99
Silva, José Celestino da, 46
Silva, Pereira da, 64, 87, 91
Silva Porto, António Francisco Ferreira da, 28
Silveira, António da, 8, 10
Simango, Uria, 156
Sinai, peace missions, 186
Sinédrio, 7–8
Skellater, John Forbes, 5
slave trade, 22, 23, 24–5, 31, 41–2
Slovenia, NATO membership, 174
Solano, General Francisco, 5
Sousa, António Teixeira de, 38
South Africa
British settlements, 29
British South African Company, 40
British trade network, 24
Cahora Bassa project, 157
claims over southern Mozambique, 75
gold mines, 40
Soviet Union
Angola decolonisation war, 151
collapse of, 167
containment of, 101
Guinea-Bissau decolonisation war, 153, 154
invasion of Afghanistan, 164–5
Portuguese UN membership application, 110
pre-constitutional government, 163
Second World War, 97, 98
as threat to Western world, 109, 111, 112, 113, 115, 173
see also Cold War; Russian Federation
Spaak, Paul-Henri, 115
Spain
Anglo-American military bases in the Azores, 99–100
annexationist lobby, 52, 56, 74, 84
Carlist wars, 14, 16
diplomatic rapprochement with Britain, 49, 52, 56, 74
Eurocorps, 169
European-peninsular thesis, 55–6
Falange, 83, 84
Franco-Spanish agreement (1912), 51
French invasion (1823), 3–4, 10
Iberian Bloc, 98, 109, 115, 120
Iberian Pact, 84, 94, 96, 99–100, 115
Indian occupation of Goa, Daman and Diu, 130

liberation wars, 4
military coup (1936), 80–1
moderate party election victory (1933), 79
monarchist incursions into Portugal, 52, 56, 64, 74
NATO exclusion, 114, 115, 116
NATO membership, 176, 177
negotiations with Germany during WWII, 96, 97
"peninsular friendship" notion, 77–8, 84, 94, 130
Peninsular Wars, 1
Popular Front election victory (1936), 79, 88
Portuguese Armed Forces rearmament, 144
Portuguese civil war (1846–1847), 20
Portuguese civil wars (1820–1834), 4, 13
Portuguese colonial policies, 133, 135
Portuguese expeditionary division (1835), 16
Quadruple Alliance (1834), 15
rapid reaction forces, 170
"red biennium" (1931–1933), 79, 88
relations with Britain in Franco era, 92–3
relations with Portugal, 51–2, 55–8, 78–80, 84, 98, 109
Republic threat to *Estado Novo*, 78–9, 80, 88
Second World War, 95–7, 98, 102
"Spanish peril" spectre, 49, 52, 56, 57, 79, 80, 88, 103
UN admission (1955), 128–9
Western European Union (WEU) membership, 169

Spanish Civil War (1936–1939), 81–5, 88
Spanish Foreign Legion, 83
Spínola, António de, 152, 153–4, 160, 162
spoliation myth, 30, 35
Stanley, Henry Morton, 28
Sukarno, 129
Sweden, rearmament of Portugal, 104
Switzerland, 50
"system of compensations", 27–8

Talleyrand, Charles Maurice de, 13
Tamagnini de Abreu e Silva, Fernando, 69
Tancos military polygon, 67–8
Tanganyika Lake, 28
Tanzania
Mozambique African National Union (MANU) founding, 148
Mozambique decolonisation war, 155, 157
National Democratic Movement of Mozambique (UDENAMO), 148

Taranco, General Francisco, 5
Teixeira de Magalhães Lacerda, Gaspar, 8
Teixeira Pinto, João, 45
Terceira, Duke of, 14, 15, 16, 17, 19
Terceira Island, 13, 40
territorial occupation principle, 29–30, 32–3
Tete, 28, 30, 155, 156, 157, 158
Third Portuguese Republic *see* Constitutional Governments (post-1976)
The Times newspaper, 29–30, 139, 158
Timor *see* East Timor; Portuguese Timor
Torres Vedras, Battle of (1846), 20
Touré, Sekou, 152, 153
Tsonga ethnic group, 40
Tungue, 33
Tunisia, 29

UDENAMO (National Democratic Movement of Mozambique), 148, 149
Ultimatum Inglês, 34–5, 39
UNAMET (United Nations Mission to East Timor), 190
UNAMI (National African Union of Independent Mozambique), 148
UNESCO, 138
UNIFIL (United Nations Interim Force in Lebanon), 192
Union of the Peoples of Angola (UPA), 131, 147, 150
Union of the Peoples of Northern Angola (UPNA), 147
UNITA (National Union for the Total Independence of Angola), 148, 151
United Nations (UN)
Angola's peace process, 186–7
anti-colonialism, 123, 129

United Nations (UN) *(continued)*
Balkans peace missions, 188–90
Chinese representation at the Security Council (1971), 126
East Timor peace mission, 190–1
founding of (1945), 110–11
Guinea-Bissau independence recognition, 154
international isolation of Portugal, 130
Lebanon peace missions, 186, 192
Mozambique decolonisation war, 158–9
Mozambique peace process, 187
Namibia peace mission, 186
non-aligned Afro-Asian movement, 129, 134
peace missions, 185–91, 192
Portugal not a founding member, 102, 110
Portugal's admission (1955), 117, 123, 128
Portuguese colonial policies, 129, 131, 134–5, 138–9, 186
Spanish admission (1955), 128–9
Western Sahara self-determination referendum, 188
United States
Angola decolonisation war, 150
Angola's peace process, 186
Azores military bases, 99, 100–1, 108, 110, 111, 113, 118, 125, 132–3, 137–8, 164–5, 173
Bush administration, 172
Clinton administration, 172
counter-guerrilla warfare, 145
defence agreement with Portugal (1957), 118, 123–4
defence agreement with Portugal (1971), 125–6
interventionism, 123
invasion of Iraq (2003), 175
Johnson administration, 134
Kennedy administration, 123, 130, 131, 132–3, 135, 136, 143, 144–5
Lajes Agreement (1948), 110, 111
Lajes supplementary defence agreement (1957), 118, 123, 132
liberation movements, 131, 137
Marshall Plan, 111
military cooperation agreement with Portugal, 111
Monroe doctrine, 50
mutual assistance agreement with Portugal (1951), 118, 121
NATO founding member, 112
NATO role, 113
NATO transatlantic link, 175
as naval power, 100, 110, 116
Nixon administration, 125, 130, 135, 137, 145
Obama administration, 175
official recognition of First Portuguese Republic (1911), 50
Portugal as NATO founding member, 115
Portuguese Armed Forces rearmament, 121, 126, 136, 137–8, 143–5
Portuguese colonial policies, 131–4, 135, 136, 137–8
post-Cold War defence policy, 172–3
relations with Constitutional Governments, 164–5, 167, 172, 173
relations with *Estado Novo*, 100–1, 111, 117–18, 121, 122, 123–4, 125–6, 128, 130, 132–4, 136, 137–8
Sakwa Plan, 123, 131–2
Salazar's distrust of, 100–1, 111, 115, 131
Santa Maria hijacking, 131
Second World War, 98–9
as sole global power, 167, 175
UN Security Council Resolution (1961) on Angola, 133, 135
Yom Kippur War, 138
see also Cold War; Vietnam War
UNOGIL (United Nations Observation Group in Lebanon), 186
UNTAET (United Nations Transnational Administration for East-Timor), 191
UNTAG (United Nations Transition Assistance Group in Namibia), 186
UPA (Union of the Peoples of Angola), 131, 147, 150
UPNA (Union of the Peoples of Northern Angola), 147
USSR *see* Soviet Union

Vasconcelos, César, 21
Vasconcelos, Loulé, 21

Vatican, Portuguese colonial policies, 133, 138
vátua State, 39, 40–1
Vau do Pembe disaster, 43
Verona, Congress of (1822), 3–4, 10
Versailles Peace Conference (1919), 42, 44, 66, 72, 74–5
Vienna, Congress of (1815), 2–3, 4
Vietnam War, 134, 141, 145, 146, 160
Vilafrancada, 10, 11
Vintista government, 8, 9, 10, 11, 15
"Viriatos", 83, 84
Volunteer Battalions of Commerce, 12

War of the Oranges (1801), 5
Warsaw Pact, 167, 174
Washington Treaty (1949), 110, 112, 116, 178
Wellington, Arthur Wellesley, Duke of, 5
West Germany
  rearmament of the Portuguese Armed Forces (1961–1974), 136, 138, 144, 145
  relations with *Estado Novo*, 134, 136, 144, 145
  see also Germany
Western European Union (WEU), 167, 168–71, 173, 175, 189
Western Sahara, peace missions, 186, 188
William IV, King of Great Britain, 4
Windsor, Treaty of (1899), 49
Wiriamu massacre, 139, 158
"wolfram question", 101, 105

Yemen, peace missions, 186
Yom Kippur War, 126, 138
Yugoslavia (former), 188–9

Zaire River, 28, 29, 31, 32, 42
Zambezi area, 22, 30, 33
Zambezi River, 28, 31, 157, 158
Zambezia, 41, 155, 158
Zambia, 148, 156
ZAMCO consortium, 157